Environmental Politics and Theory

Series Editors

The premise of this series is that the current environmental crisis cannot be solved by technological innovation alone. The environmental challenges we face today are, at their root, political crises involving political values, institutions and struggles for power. Therefore, environmental politics and theory are of the utmost social significance.

Growing public consciousness of the environmental crisis and its human and more-than-human impacts, exemplified by the worldwide urgency and political activity associated with the problem and consequences of climate and earth system change make it imperative to design and achieve a sustainable and socially just society.

The series publishes inter- and multi-disciplinary scholarship that extends the theoretical dimensions of green political theory, international relations, philosophy, and earth system governance. It addresses the need for social change away from the hegemonic consumer capitalist society to realize environmental sustainability and social justice.

Duncan Weaver

The Aarhus Convention

Towards Environmental Solidarisation

palgrave
macmillan

Duncan Weaver
Social Sciences and Humanities
University of Suffolk
Ipswich, UK

ISSN 2731-670X ISSN 2731-6718 (electronic)
Environmental Politics and Theory
ISBN 978-3-031-43535-5 ISBN 978-3-031-43536-2 (eBook)
https://doi.org/10.1007/978-3-031-43536-2

Cover illustration: Evandro Maroni/Stockimo/Alamy Stock Photo

This Palgrave Macmillan imprint is published by the registered company Springer Nature Switzerland AG
The registered company address is: Gewerbestrasse 11, 6330 Cham, Switzerland

Acknowledgements

In a recent meeting, a university leader reminded their audience that "to work in a university is a privilege". I nodded silently. To learn, and pass knowledge on, is no luxury. Nor is it something to be taken for granted. It is an immense privilege. I recognize it, and I am deeply thankful for it.

The thoughts lodged in this book would not have arisen without the generosity of Keele University, whose Vice Chancellor's Scholarship and Graduate Teaching Assistantship enabled six years' study without impediment. Neither would they have arisen without the generosity of the Department of International Politics, Aberystwyth, whose David Davies Scholarship afforded me a year out of Keele to study under, and with, the loveliest people.

I thank my Keele doctoral supervisors, Dave Scrivener and John Vogler, for their mentorship, challenge and leadership by example. I chose Keele for my undergraduate studies because it was enjoying a boom in green politics and sustainability, both domestic and international: a boom that continues today. Thanks go to internal examiner Phil Catney (Keele) and external examiner Hugh Dyer (Leeds). None of these four chaps abandoned me during my brief escape from education. And for that, I am grateful.

Jonathan Gilmore (Manchester), Ben Stanford (Liverpool John Moores), Femi Awopegba (Manitoba), Matt Gallagher (City College Norwich), Kaja Weaver (City College Norwich) and David James (Suffolk) critiqued portions of drafts, giving much to reflect on. Robert

Falkner (LSE) kindly discussed the greening of the English School during my doctorate; I thank him for having signposted me towards literature in this project. I thank Matt Weinert (Delaware) for his insights during and after my studies. I thank Avi Brisman (Eastern Kentucky) for permitting a discussion of the violence of silence during my studies. I recently had the good fortune to present some of the book's logic at the *Together for Transformation* conference at Suffolk, and also to pitch some of my thoughts about environmental harm and political theory to Nigel South (Essex). The chief theoretical findings of this book were pitched to the International Society section of the EISA Pan European Conference 2023, Universität Potsdam, and I express thanks for colleagues' incisive comments there. Ray Weaver kindly proofread the entire work, and I am very grateful for Gor Samvel having read and critiqued the manuscript from a legal vantage point.

But all that said: I alone remain fully responsible for all and any insufficiencies.

I thank my doctoral participants for shedding light on Aarhus, and showing me the human side of international environmental politics and law. The Programme Management Unit, United Nations Economic Commission for Europe, was generous enough to host me for half a year after my doctorate. What a rare chance to 'live' international relations, and to meet such an inspiring bunch of folks. Never did I see more mammoth a body of work undertaken, with such efficiency and good humour, than at the Palais.

But ultimately, none of this study gets done without family. I thank my parents for their total, unquestioning support. I thank my Duplje family for their tolerance whilst I typed in their roof and kitchen. I was quieter than I would have wished, and can only hope to make up for lost time over čevapčiči. But most of all, I thank Kaja, who encouraged me without the slightest reservation in this book, and who supported me all the way.

Ljubim te, moja pustolovka!

Contents

1 Introduction 1
Introducing Aarhus 1
Key Argumentation 8
Structure 10
Bibliography 14

2 Theoretical Framework 19
Introduction 19
Realism 23
Revolutionism 25
Rationalism 31
Pluralism 33
Solidarism and Solidarisation 36
Conclusion 40
Bibliography 41

3 Propagation: The Aarhus Convention's International Context 49
Introduction 49
International Legal and Diplomatic Context 51
Geopolitical Context 69
Environment for Europe 77
Conclusion 87
Bibliography 91

4 Germination: The Aarhus Convention's Procedural Trinity 105
Introduction 105
The First Pillar: Access to Information 111
The Second Pillar: Public Participation 119
The Third Pillar: Access to Justice 127
Conclusion 132
Bibliography 134

5 Growth: The Aarhus Convention's Organisational Infrastructure 141
Introduction 141
The Strategic Tier: The Meeting of the Parties 143
Operational Tier: Working Group of the Parties, Bureau, Task Forces and Secretariat 157
The Tactical Tier: Aarhus' Compliance Mechanism 159
Conclusion 168
Bibliography 170

6 Conclusion: Towards Solidarisation 179
Introduction 179
Realism 180
Revolutionism 182
Rationalism: Towards Solidarisation 184
Bibliography 190

Index 193

About the Author

Duncan Weaver is Senior Lecturer at University of Suffolk, United Kingdom. Previously, he was Senior Instructor (Anglia Ruskin), Lecturer (City College Norwich) and Graduate Teaching Assistant (Keele). He studies two related problems in domestic and international society: (a) the need for state governance to be more 'humankind'; and (b) the geopolitical susceptibilities of marginal spaces and actors. His concerns sit between critical geopolitics and the English School of IR. His work was placed in *Geopolitics, International Environmental Agreements*, *Social Movement Studies*, *Oxford Research Encyclopedia of International Studies* and *Comparative Civilizations Review*.

Abbreviations

BRI	Belt and Road Initiative
CBD	Convention on Biological Diversity
COP	Conference of the Parties
CSCE	Conference on Security and Cooperation in Europe
EC	European Community
ECHR	European Convention on Human Rights
ECJ	European Court of Justice
ECLAC	Economic Commission for Latin America and the Caribbean
ECO	Environmental Citizens' Organisation
ECtHR	European Court of Human Rights
EDI	Equity, Diversity and Inclusion
EEB	European Environmental Bureau
EECCA	Eastern Europe, Caucasus and Central Asia
EIA	Environmental Impact Assessment
EU	European Union
FSB	Federal Security Service of the Russian Federation
GMOs	Genetically Modified Organisms
HE	Higher Education
IAEA	International Atomic Energy Agency
ICC	International Criminal Court
IGO	Intergovernmental Organisation
IR	International Relations
IUCN	International Union for the Conservation of Nature
MDGs	Millennium Development Goals
MEA	Multilateral Environmental Agreement
MOP	Meeting of the Parties

NATO	North Atlantic Treaty Organisation
NGO	Nongovernmental Organisation
NIR	National Implementation Reporting
OSCE	Organisation for Security and Cooperation in Europe
R2P	Responsibility to Protect
SDGs	Sustainable Development Goals
UK	United Kingdom
UN	United Nations
UNCED	United Nations Conference on Environment and Development
UNCHE	United Nations Conference on the Human Environment
UNECE	United Nations Economic Commission for Europe
UNEP	United Nations Environment Programme
UNESCO	United Nations Educational, Scientific and Cultural Organisation
UNFCCC	United Nations Framework Convention on Climate Change
US	United States of America
USSR	Union of Soviet Socialist Republics
WCED	World Commission on Environment and Development

CHAPTER 1

Introduction

Introducing Aarhus[I]

International environmental agreements (see Andresen et al., 2011 for a comprehensive introduction) are like cultivated flowering plants. They are propagated; they germinate; sometimes they grow. Sometimes, they reproduce. Some agreements succeed, whilst others are less lucky. Nonetheless, each agreement provides its broader ecosystem—both its human and nonhuman constituencies—with benefit. Benefits are not only discretely environmental; they have broader impact on the human and societal constitution: on health, well-being, life quality and sometimes—as with the Aarhus Convention—on human empowerment and democracy. But cultivated flowering plants are controlled, as any gardener like me is aware, to prevent them growing too wild. They tend not to "go rogu" and surpass one's requirements. They are cultivated by the gardener, for the gardener and in the gardener's interests. It is the gardener who controls them. Any growth is at their discretion. *They* choose the extent to which the plant develops. If they for some reason are not inclined to continue supporting its growth, they will act accordingly. As such, the plant enriches the garden and gardener to the extent permitted.

This metaphor—and I admit it is a quaint one—offers an entrance into this book's key position. On one hand, international environmental agreements enrich their "gardeners", the consenting states that negotiated, signed, ratified and implemented them. The agreements' legal

D. Weaver, *The Aarhus Convention*, Environmental Politics and Theory, https://doi.org/10.1007/978-3-031-43536-2_1

provisions, implementation regimes and compliance activities enrich and mature state sovereignty. When a state joins such an agreement, it must "work more" to remain a legitimate member of international society, and to ensure its sovereignty—the authority and legitimacy with which it governs domestically and exerts actorhood internationally—continues to be respected by (a) its population, (b) its state counterparts and (c) global civil society. I include in the latter, inter alia, nongovernmental organisations (NGOs), social movements, loose human associations and networks and individuals acting within and across borders (see Wapner, 1996). States invariably adapt, to some lesser or greater extent, their domestic behaviours to ensure compliance with an agreement (for Aarhus' binding requirement, see Aarhus Convention 1998, Article 3.1). As such, the agreement—when implemented in the best of faith by consenting states—yields domestic impact, "intruding" into sovereignty. And it also goes without saying that such agreements aim to enrich, enhance and ultimately protect different elements of the literal ecosystem. This is their goal: to improve the ecosystem *within*, and *beyond*, state borders. As such, when a group of states, mutually concerned for a shared environmental issue, reach legal agreement on rules and procedures to address that issue, they form an environmentally oriented international society, marked by different thresholds of responsibility than existed before the agreement's codification. One would anticipate that those thresholds are stricter and more ambitious. But sometimes they will also represent the less satisfactory "lowest common denominator" outcomes of arduous negotiations between states whose interests conflict and compete.

On the other hand, however, it is not just the agreement that regulates its consenting states. Above, I alluded to the gardener ensuring that the plant does not "outgrow" its welcome. States themselves regulate the agreement's development. As we will learn in this book, states continuously affirm and iterate the sanctity of their sovereignty, and the salience of their sovereign prerogatives—to protect, promote and pursue their *particular* and diverse national interests—throughout their participation in such an international environmental agreement. Such iterations of sovereignty occur in order to (a) prevent usurpations of sovereignty, (b) ensure that potentially radical or transformative progress in IR remains within states' threshold of feasibility and tolerance and (c) ensure the relative stability of international order. Only if these circumstances are achieved will statespeople feel sufficiently comfortable taking tentative steps of ***cautious incrementalism*** in international environmental politics and law: steps towards more ethically ambitious progress.

As this book will demonstrate, some states express reservation, pursuant to classical international law, when they perceive an agreement's provisions to surpass their threshold of feasibility and tolerance. Some states refuse to negotiate an agreement, despite their eligibility to do so. Others negotiate but avoid signing it. Yet others withdraw from an agreement entirely, despite having been Party to it. Sovereignty remains, without any modicum of doubt, the currency spent by states on the diplomatic "pitch" of IR. As this book will show, states can and do unilaterally leave the "pitch" when they deem their ***particular*** interests no longer served, or when they deem those interests outright hindered.

As such, mutual interdependence exists between "gardener" and "plant", between state and international agreement. Both regulate one another. Both are co-constitutive. This is a key logic animating the book. Suffice it to note here that the international agreement for which I am discretely concerned is the Aarhus Convention. Aarhus is a particularly unique and impressive flower in the garden of multilateralism. Aarhus was "one of the major results of the Rio Declaration" (United Nations, 2011: 1), the latter a key outcome of the United Nations Conference on Environment and Development (UNCED), held in Rio de Janeiro in 1992 (United Nations General Assembly, 1992). Aarhus opened for signature at the fourth ministerial conference of the Environment for Europe initiative, Aarhus, Denmark, on 25 June 1998 (UNECE, 1998). It entered into force on 30 October 2001. When it opened, UN Secretary-General Kofi Annan deemed Aarhus "a milestone in increasing transparency and participatory possibilities in the environmental arena. It will help put an end to environmental secrecy. It will promote democracy and protect the right of everyone to live in a healthy environment" (UNECE, 1998: 4).

Indeed, in the quarter century since it opened for signature, Aarhus received praise as a pioneer agreement that helped lead the development of environmental democracy and environmental rights (Day, 2015: 185; Dobson, 2003: 149–150; Eckersley, 2004: 250–251; Jendrośka, 2009: 496; 2012: 73; Kingston et al., 2022: 2; Pallemaerts, 2011a: 3; Pedersen, 2008: 93). However, whilst Aarhus has been applauded in the highest echelons of world affairs, and by the most esteemed lawyers and political commentators, there was only modest engagement with its ethical implications. Until this book, I had only scratched the surface of the ethical momentum that Aarhus had built (Weaver, 2018). The question remained as to exactly what consequences the Convention yielded for ethical progress in a world that continues to be prevailed upon by states. This is the task that the book assumes.

Before setting to it, I should note that Aarhus has 46 state Parties[2] plus the European Union,[3] whose 27 member states constitute circa 59% of Party membership (United Nations Treaty Collection, 2023; see UNECE, 2023). It should be noted that Belarus recently expressed its intention to withdraw from the Convention, and this development will be revisited later in the book. Employing UN (2020: 165) classifications, 31 Parties (circa 67%) are developed economies,[4] 14 (circa 30%) are economies in transition[5] and 1 (circa 2%), Guinea-Bissau in West Africa (see Etemire, 2023: 186), is a developing economy. 20 Parties (circa 43%) or their historical predecessors were associated with the Warsaw Pact.[6] 28 (circa 61%) are members of the North Atlantic Treaty Organisation (NATO).[7]

We might agree, then, that Aarhus is implicated in democratisation, predominantly but not exhaustively in the UNECE region. Its implication in democratisation centres on its attempted amplification of human environmental "voice": the enhancement of the human capacity to "hear and be heard" in potentially harmful environmental decision-making procedures. To achieve this, Aarhus assigns a trinity of procedural rights to information access, public participation and access to justice "without discrimination as to citizenship, nationality or domicile" (Aarhus Convention 1998, Article 3.9). To employ my earlier analogy, the ethical intent is to better empower the gardeners to manage the ecosystem. By enlisting an active, environmentally minded public, and by stimulating a co-creative dialogue between state and human actors, the gardeners should be better placed to ensure that not only decision-making *procedures* but also *outcomes* are more just and more substantively benign to the environment.

In environmental contexts, then, Aarhus more firmly and justiciably renders states responsible for the humans and environments occupying their jurisdictions. As this book will show, the Convention helps to "tune up" human participatory capacity in pursuit of enhanced accountability-holding in environmental governance. It codifies stricter provisions for environmental accountability than existed before it. This benefits the human as referent of ethical concern, and the environment as ultimate beneficiary of more reasoned, and environmentally sensitive decision outcomes.

Far be it from me to claim that such laudable rationales are always borne in reality. As the book gains traction, we will witness a number of potential obstacles, not least relating to the persistence of state consent, a pluralist mainstay (Hurrell, 2007: 49) denoting the ability and willingness of states to leave the diplomatic "pitch" at will. But what I *do* assert is that the international society, constituted by Aarhus' Parties, was challenged to be just that little more responsible, environmentally conscious, "humankind", when Aarhus opened for signature a quarter century ago. I have no qualms arguing here that the world was better off with, rather than without, Aarhus' emergence.

Having made some opening remarks on the ethical observations to come, it will be helpful for the reader if I introduce the structure of the Convention. A key focus, in this book, is on Aarhus' trinity of procedural rights. The strongest evidence of Aarhus' ethical traction is found in its procedural provisions for information access, public participation and access to justice. As such, Chapter 4 of this book is devoted to the procedural trinity, which is hence only mentioned here. Let us start with the Convention's opening preamble, which declares Aarhus' aspirations. Three points need to be made about it; they relate to the nonbinding recognition of a substantive environmental right; the interdependence of human and environmental rights and responsibilities, and Aarhus' role as a democratic as well as environmental pioneer. I will address each in turn.

Firstly, the preamble (Aarhus Convention 1998) recognises a nonbinding substantive right to environment, affirming "the need to protect, preserve and improve the state of the environment and to ensure sustainable and environmentally sound development". More explicitly, it (Aarhus Convention 1998, preamble) notes "that every person has the right to live in an environment adequate to his or her health and well-being, and the duty, both individually and in association with others, to protect and improve the environment for the benefit of present and future generations". I must stress that Aarhus' subsequent operative text assigns binding *procedural* rights, not *substantive* environmental rights. Recognition of a substantive right remains preambular aspiration. This is reflective of broader difficulties to codify a substantive environmental right in international law (Cramer, 2009; Shelton, 2006, 2010). To overcome such difficulties, procedural rights are used to safeguard the due process of "good governance", optimising the likelihood of decision-making *outcomes* being environmentally benign, by ensuring that the *processes* leading to them are as just as they can be. By codifying and enforcing such procedural rights as those to information access, public participation and access to justice, any decision-making improprieties should be more visible and manageable in terms of redress. The hope is that procedural rights optimise human capacities to understand and influence decisions that may harm them and their environments (Dellinger, 2012; Getliffe, 2002; Okowa, 1997). They should help "amplify" human "voice" in environmental decision-making (see Cox, 2006; Depoe et al., 2004), reducing the risks of an environmental "violence of silence" ensuing (Brisman, 2013; see Linklater, 2011: 99; Shapcott, 2008: 198).

Secondly, the preamble casts human and environmental rights and responsibilities as interdependent. The logic is that there can be no enjoyment of discretely environmental rights without human rights, and vice versa. Likewise, there can be no enjoyment of rights without the undertaking of responsibilities, on the part of humans and their associations. This point should not be understood as some form of neoliberal "offloading" of duty by states onto humans. It is, rather, deeply empowering. The theme of empowerment ebbs throughout this book, and readers will become familiar with what I term the *cosmopolitan impulse*, which captures Aarhus' moral cosmopolitan ethical drivers of human empowerment, voice, state-human co-creation and "humankind" governance modalities. In this regard, the preamble (Aarhus Convention 1998) notes "that adequate protection of the environment is essential to human well-being and the enjoyment of basic human rights, including the right to life itself". It (Aarhus Convention 1998, preamble) considers also that the nonbinding substantive right, and associated duty, is contingent on procedural rights. There is acknowledgement that to discharge environmental duties, humans must be endowed with environmental awareness and education (Aarhus Convention 1998, Article 3.3) and furnished with information that, after all, is held by public authorities "in the public interest" (Aarhus Convention 1998, preamble).

Thirdly, it is clear from the preamble that Aarhus is a deliberate innovation in *democracy*, as much as it is an *environmental* instrument (see in particular Chapter 3). It evinces awareness of the intrinsic as well as instrumental benefits of information access, public participation and access to justice for humans and their associations. It aims to develop accountability and transparency, recognising the "desirability of transparency in all branches of government" and the roles played by humans, NGOs and private sector actors in environmental protection (Aarhus Convention 1998, preamble). It (Aarhus Convention 1998, preamble;

emphasis added) is no less than "*Convinced* that the implementation of this Convention will *contribute to strengthening democracy*" in the UNECE region.

Aarhus' legally operative text begins after the preamble. Article 1 sets the Convention's objective. This is as close as Aarhus gets to codifying a substantive environmental right, but I repeat that Article 1 does not constitute legally binding codification of a substantive right. It (Aarhus Convention 1998, Article 1; emphasis added) states that "*In order to contribute to* the protection of the right of every person of present and future generations to live in an environment adequate to his or her health and well-being, each Party shall guarantee the rights of access to information, public participation in decision-making, and access to justice in environmental matters in accordance with the provisions of this Convention". "Shall" implies binding duty; thus the *procedural* rights are binding, not the protection of environmental adequacy. That said, the provision has indeed been deemed pioneering (Brady, 1998: 70; Wates, 2005: 2).

Article 2 sets the Convention's definitions. Parties[8] and public authorities[9] are defined. A broad approach to environmental information is adopted (see Chapter 4); the *public* is distinguished from the *public concerned* (see Chapter 4). Article 3 contains general provisions governing Aarhus' application. The Convention's provisions are cast as a floor, not a ceiling (Aarhus Convention 1998, Article 3.5). Parties are obliged to ensure that humans are assisted in exercising their Aarhus rights (Aarhus Convention 1998, Article 3.2). Parties are obliged to recognise and support civil society (Aarhus Convention 1998, Article 3.4), and they are mandated to promote Aarhus' principles in international forums other than the Convention itself (Aarhus Convention 1998, Article 3.7). It is worth reiterating, at this juncture, the significance of the provision that Aarhus' rights apply "without discrimination as to citizenship, nationality or domicile and, in the case of a legal person, without discrimination as to where it has its registered seat or an effective centre of its activities" (Aarhus Convention 1998, Article 3.9). This helps nullify unhelpful distinctions of nationals from non-nationals, in the provision of the rights. It is consistent with the cosmopolitan premise that one human life (or at least one instance of human agency) is equal in worth to another. Such "humankindness" is evident in Article 3.8's (Aarhus Convention 1998) requirement for Parties to ensure that persons exercising their rights "shall not be penalised, persecuted or harassed in any way".

Thereafter, the trinity of procedural rights is provided. The first pillar, information access, is codified in Articles 4 and 5. The second, public participation, is provided in Articles 6, 7 and 8. The third, access to justice, is found in Article 9. What remains, in the Convention text, provides Aarhus' "working parts": its Meeting of the Parties (MOP) (Article 10), voting provisions (Article 11), secretariat (Article 12), annexes (Article 13), amendment arrangements (Article 14), compliance mechanism (Article 15), dispute settlement provisions (Article 16), signature (Article 17) and subsequent entrance into force (Article 20), depositary (Article 18), arrangements for ratification, acceptance, approval and accession (Article 19), arrangements for Parties' withdrawal (Article 21) and authentic texts (Article 22). Two annexes come at the end. The first non-exhaustively enumerates the specific environment-related activities discussed in Article 6 participatory provisions. The second offers particulars governing the arbitration process.

Key Argumentation

What we see, from the above, is an apparently mainline international environmental agreement, containing the machineries that scholars of environmental multilateralism would expect to find. Yet Aarhus is driven, as I noted, by a *cosmopolitan impulse* that infuses the traditions of IR—state consent, sovereignty, the sovereign prerogative of pursuing national interests in orthodox legal and diplomatic channels—with human-oriented ambition. Aarhus evinces the possibility, I argue, of taking IR's orthodoxies and enriching them. Such a cosmopolitan impulse, which was regarded by a scholar of the International Criminal Court (ICC) (Roach, 2005: 475) as cosmopolitan intent, ethically inclines states towards the vulnerability of humankind as a whole, as opposed to the vulnerability of solely their counterpart states or domestic citizenries in international society.

To explicate this within a compelling framework, I propose that Aarhus demonstrates the possibility and appeal of *solidarisation* (Ahrens, 2017, 2018, 2019; Ahrens & Diez, 2015) in environmental contexts. Let me not spoil the next chapter, but rather note that this idea is rooted in the English School of IR (see in the first instance Buzan, 2004, 2014, 2018; Falkner, 2021; Falkner & Buzan, 2019, 2022; Hurrell, 2007; Linklater & Suganami, 2006; Navari, 2021a, 2021b). The English School is often associated with a preference for *international society*. I define an international society as a deliberate convergence of states upon shared issues of concern, involving states' conscious multilateral investment of sovereignty in measures to address those shared issues. Two understandings of international society are elicited from the English School literature (Linklater & Suganami, 2006: 199–206; Weinert, 2011). On one hand, a "thin" pluralism foregrounds states' pursuit of minimal coexistence, seeking to (a) preserve their sovereign prerogatives, (b) maintain the diversity of particular sovereign jurisdictions and interests in IR and (c) prevent harm being caused and sustained by states. Such pluralism is premised on the notion that diversity is a "reality of world politics", and that international order would be compromised by those attempting to silence, escape from, impose upon or universalise over diversity (Jackson, 2003: 403). The logic is that to silence diversity is to destabilise international order itself, thus preventing any possibility of manoeuvring towards more ethically exacting progress.

On the other hand, a "thicker" solidarism (see Wheeler, 1996, 2000; Wheeler & Dunne, 1996) continues to foreground state sovereignty, but seeks closer convergence between states, the mobilisation of sovereignty as a more ethically robust "force for good", and—crucially for this present study—the inclination of states' ethical priorities towards human as opposed to solely state-oriented vulnerabilities. The reader will become familiar with my use of the term "humankindness", which entails adopting a moral cosmopolitan ethic whilst retaining the traditional power frameworks of sovereignty.

With this latter point in mind, *solidarisation* is a relatively fresh take on this pluralist–solidarist dynamic. It is a process whereby solidarist, potentially radical human-oriented ethical rationales are contained and confined by pluralist, inherently statist power frameworks. Under solidarisation, pluralism and solidarism need not be "either/or" phenomena. They can mutually coexist and reinforce each other. Aarhus, I argue, evinces the feasibility and benefit of fostering solidarist "humankind" values inside existing pluralist arrangements. Aarhus demonstrates the possibility of solidarist values concomitantly (a) ethically agitating sovereignty whilst (b) being contained and confined by the strictures of sovereignty. This echoes Ahrens' (2018: 7) finding that under solidarisation, progress "can hardly be radical but…needs to connect with and take into consideration previously existing structures". As such, post-Westphalian transformation is not elicited from my study of Aarhus. Any potentially radical progress that *is* achieved adheres to a more realistic cosmopolitanism

(Beck & Sznaider, 2006), rather than an emancipatory unshackling of humans from sovereignty. Aarhus indicates that solidarism can and should contribute to an ethical maturation of sovereignty, in environmental contexts, without usurping it. The finding throughout is that sovereignty will regulate, confine and contain the growth of solidarism, in order to preserve the stability of international order and states' coexistence. What follows is an observation of ethical ambition being garnered in a way that (a) renders sovereignty more responsible, and more acutely attunes states to human and environmental vulnerability, whilst (b) being *counterbalanced* by iterations of sovereignty that are inevitable in, and crucial to, the maintenance of international order. In sum, my key argumentation is that solidarist ethical progress will only be as radical as consenting states allow. It will remain accompanied by (a) states' iterations of the sanctity of their sovereignty and (b) attempts to prevent usurpations of sovereignty by "runaway" transformation. Counterbalancing in favour of sovereign prerogatives is used to mitigate such risks, and in order to ensure the relative stabilisation of international order.

Structure

What follows is not another bookend on the lawyer's shelf (see Banner, 2015; Barritt, 2022; Pallemaerts, 2011b). There is no shortage of recent and excellent legal scholarship on Aarhus (Braig et al., 2022; Fasoli & McGlone, 2018; Hedemann-Robinson, 2022; Lee, 2023; Peters, 2018; Ryall, 2023; Samvel, 2020; Smyth, 2022). My intention, as someone discretely concerned for ethical progress in IR, is to focus on Aarhus as a radical additive to residually statist environmental multilateralism. I consciously treat the Convention from an ethically minded IR vantage point, donning English School lenses to elicit, from the archives, the degree of (a) solidarist potential and (b) pluralist persistence in Aarhus' propagation, germination and growth. It is through this task that the possibilities for solidarisation in environmental contexts are elicited.

With this in mind, Chapter 2 offers an English School theoretical framework drawing on its historical continuum (Wight, 1991), with realist international system and revolutionist cosmopolis occupying the two ends, and rationalist international society roughly equidistant therebetween. Each element of the continuum offers an imaginary of IR against which to assess "real world" developments such as international environmental agreements. International society is portrayed as a prudent, cautious middle way, mindful of the realities of anarchy, the absence of world government, without negating the possibility of propagating lesser (pluralist) or greater (solidarist) degrees of progress. Solidarisation is elucidated as a process in which solidarist progress can be embedded within existing pluralist power arrangements.

Chapter 3 explores Aarhus' *propagation*, examining the legal, diplomatic and geopolitical context in which it could be successfully proposed, negotiated and agreed. The post-1945 conditions enabling Aarhus' emergence are evaluated. First, Aarhus' legal and diplomatic context, vis-à-vis human rights and environmental protection, is seen to set the stage on which Aarhus could be proposed. Second, the geopolitical context of Aarhus' emergence is seen to be inseparable from (post-)Cold War democratisation. Aarhus would not have emerged as it did without the political recalibrations accompanying the Cold War's end. After accounting for those contexts, the chapter assesses Aarhus' emergence under Environment for Europe, a framework that welcomed human involvement in multilateralism. It is concluded that whilst a gratifying degree of solidarist ambition was evident in those contexts, it was nonetheless tempered by state consent. The chief focus, after 1945, was on securing an international order that would avoid global war and foster as nonviolent a coexistence between states as possible. There was a sense that human and environmental justice could only be pursued in the framework of stable order. Whilst human rights and environmental protection were fixed on the international agenda by the mid-late twentieth century, affording favourable conditions for Aarhus' emergence, "green" solidarist concerns were contained inside pluralist architectures that sought to safeguard state coexistence in the aftermath of global war.

Chapter 4 assesses Aarhus' *germination*, namely the provision of its trinity of procedural rights. Whilst one might assume that the rights herald revolutionary progress, what is actually observed is a mutual counterbalancing between (a) Aarhus' cosmopolitan impulse and (b) the *persistence of*, and *insistence upon*, sovereignty. Such counterbalancing indicates solidarisation, as the trinity demonstrates a suite of checks and balances that regulate and limit solidarist ambition *as well as* the predominance of sovereignty over decision-making procedures. Potentially radical progress is contained by, and confined within, the threshold of states' tolerance. The chapter gauges from the trinity a spirit of co-creation; an operationalisation of the ancient "all-affected principle" (Warren, 2017) whose dictum is "what touches all should be heard and approved by all";

an amplification of human voice and a sharpening of the Convention's legal "teeth" of enforceability and justiciability. Aarhus is seen to ethically agitate sovereignty: but the seeds of radical change are only allowed to germinate to the extent permitted by states.

Chapter 5 addresses Aarhus' *growth*, in terms of the development of its organisational infrastructure. The latter is partitioned into strategic, operational and tactical tiers. The strategic tier concerns agenda-setting, prioritisation and leadership; the Meeting of the Parties (MOP) is assessed there. Little digression from statist multilateralism is gauged. Counterbalancing in favour of states is identified, given indications of unilateralism, lacklustre national implementation reporting (NIR) practices, and the MOP having been driven to initiate a rapid response mechanism to protect environmental defenders *inside* Party jurisdictions as well as beyond.

The operational tier encompasses the "vital organs" sustaining Aarhus' life: the Working Group of the Parties, secretariat, bureau and task forces. Operationally, Aarhus is a mainline agreement whose organs sustain its constitution and ensure its continuity. However, it is in Aarhus' tactical tier, the compliance mechanism, that gratifying solidarist potential is observed. Three elements of the mechanism are found to be especially instructive: its composition, quasi-judicial working practices and entrenchment of human empowerment, given the ability of humans and NGOs to directly lodge communications of alleged non-compliance, thus initiating proceedings that can result in domestic change. Solidarist potential is identified in the mechanism pushing Aarhus towards a trajectory of environmental *actio popularis*. But such potential is tempered by a suite of checks that serve to counterbalance the arrangements in favour of states, preventing usurpation of sovereignty and perpetuating the state as lead actor.

These findings are distilled in a conclusion, in Chapter 6, that deems Aarhus a beacon of solidarist potential in a residually inter*national* society. Lessons should be learned from Aarhus, a quarter century after it opened for signature. They relate to the viability and worth of propagating "humankindness" in the existing power frameworks of IR. They concern the benefits of safeguarding human empowerment, amplifying human voice and embracing co-creation in environmental multilateralism. They remind us that, rather than seeking post-Westphalian transformation, benefits can be gained from enriching IR's extant power arrangements; from institutionalising a responsible sovereignty that can spearhead ethical progress whilst maintaining international order and from recognising that mutual interplay between human and sovereign prerogatives can (a) help ethically mature sovereignty whilst (b) reinforcing sovereignty's capacity to delimit transformative progress, thus safeguarding as safe and orderly an international coexistence as possible.

Notes

1. For brevity I will reference, within the prose, the Convention on Access to Information, Public Participation in Decision-making and Access to Justice in Environmental Matters simply as "Aarhus Convention 1998".
2. The Aarhus Convention's 46 state Parties, notwithstanding the European Union, are: Albania, Armenia, Austria, Azerbaijan, Belgium, Bosnia and Herzegovina, Bulgaria, Croatia, Cyprus, Czechia, Denmark, Estonia, Finland, France, Georgia, Germany, Greece, Guinea-Bissau, Hungary, Iceland, Ireland, Italy, Kazakhstan, Kyrgyzstan, Latvia, Lithuania, Luxembourg, Malta, Montenegro, Netherlands, North Macedonia, Norway, Poland, Portugal, Moldova, Romania, Serbia, Slovakia, Slovenia, Spain, Sweden, Switzerland, Tajikistan, Turkmenistan, Ukraine and UK. This accounts for Belarus' withdrawal.
3. For the purpose of the Aarhus Convention's entrance into force, the European Union's deposition of its ratification instrument does not count in addition to those of its member states (United Nations Treaty Collection, 2023). European Union member states are: Austria, Belgium, Bulgaria, Croatia, Cyprus, Czechia, Denmark, Estonia, Finland, France, Germany, Greece, Hungary, Ireland, Italy, Latvia, Lithuania, Luxembourg, Malta, Netherlands, Poland, Portugal, Romania, Slovakia, Slovenia, Spain and Sweden.
4. The 31 developed economy Parties are: Austria, Belgium, Bulgaria, Croatia, Cyprus, Czechia, Denmark, Estonia, Finland, France, Germany, Greece, Hungary, Iceland, Ireland, Italy, Latvia, Lithuania, Luxembourg, Malta, Netherlands, Norway, Poland, Portugal, Romania, Slovakia, Slovenia, Spain, Sweden, Switzerland and UK.
5. The 14 economies in transition are: Albania, Armenia, Azerbaijan, Bosnia and Herzegovina, Georgia, Kazakhstan, Kyrgyzstan, Moldova, Montenegro, North Macedonia, Serbia, Tajikistan, Turkmenistan and Ukraine.
6. The 20 Parties, or their predecessors, with some historical roots in the Warsaw Pact are: Albania, Armenia, Azerbaijan, Bulgaria, Czechia, Estonia, Georgia, Germany, Hungary, Kazakhstan, Kyrgyzstan, Latvia, Lithuania, Moldova, Poland, Romania, Slovakia, Tajikistan, Turkmenistan and Ukraine.

7. The 28 NATO member Parties are: Albania, Belgium, Bulgaria, Croatia, Czechia, Denmark, Estonia, Finland, France, Germany, Greece, Hungary, Iceland, Italy, Latvia, Lithuania, Luxembourg, Montenegro, North Macedonia, Netherlands, Norway, Poland, Portugal, Romania, Slovakia, Slovenia, Spain and UK.
8. Under Article 2.1 (Convention on Access to Information, Public Participation in Decision-making and Access to Justice in Environmental Matters, 1998), a Party is defined as "a contracting Party to this Convention".
9. Under Article 2.2 (Convention on Access to Information, Public Participation in Decision-making and Access to Justice in Environmental Matters, 1998), a public authority is "(a) Government at national, regional and other level; (b) Natural or legal persons performing public administrative functions under national law, including specific duties, activities or services in relation to the environment; (c) Any other natural or legal persons having public responsibilities or functions, or providing public services, in relation to the environment, under the control of a body or person falling within subparagraphs (a) or (b) above; (d) The institutions of any regional economic integration organisation referred to in Article 17 which is Party to this Convention. This definition does not include bodies or institutions acting in a judicial or legislative capacity".

Bibliography

Ahrens, B. (2017). *The solidarisation of international society: The EU in the global climate change regime* (GLOBUS Research Paper 5/2017). University of Oslo.

Ahrens, B. (2018). *AmbigEUity: The EU and the solidarisation of international society* (PhD thesis). Eberhard Karls Universität Tübingen.

Ahrens, B. (2019). The European Union between solidarist change and pluralist re-enactment. In T. Brems Knudsen & C. Navari (Eds.), *International organisation in the anarchical society: The institutional structure of world order* (pp. 265–292). Palgrave Macmillan.

Ahrens, B., & Diez, T. (2015). Solidarisation and its limits: The EU and the transformation of international society. *Global Discourse, 5*(3), 341–355.

Andresen, S., Boasson, E., & Hønneland, G. (Eds.). (2011). *International environmental agreements: An introduction*. Routledge.

Banner, C. (Ed.). (2015). *The Aarhus Convention: A guide for UK lawyers*. Hart.

Barritt, E. (2022). *The foundations of the Aarhus Convention*. Hart.

Beck, U., & Sznaider, N. (2006). Unpacking cosmopolitanism for the social sciences: A research agenda. *British Journal of Sociology, 57*(1), 1–23.

Brady, K. (1998). New convention on access to information and public participation in environmental matters. *Environmental Policy and Law, 28*(2), 69–75.

Braig, K. F., Kutepova, N., & Vouleli, V. (2022). Playing second fiddle to the Aarhus Convention: Why the ECtHR can and should go further. *Journal for European Environmental and Planning Law, 19*, 74–102.

Brisman, A. (2013). The violence of silence: Some reflections on access to information, public participation in decisionmaking, and access to justice in matters concerning the environment. *Crime, Law and Social Change, 59*, 291–303.

Buzan, B. (2004). *From international to world society? English school theory and the social structure of globalisation*. Cambridge University Press.

Buzan, B. (2014). *An introduction to the English school of international relations*. Polity.

Buzan, B. (2018). Revisiting world society. *International Politics, 55*, 125–140.

Convention on Access to Information, Public Participation in Decision-making and Access to Justice in Environmental Matters. (1998). https://unece.org/DAM/env/pp/documents/cep43e.pdf. Accessed 9 March 2023.

Cox, R. (2006). *Environmental communication and the public sphere*. Sage.

Cramer, B. W. (2009). The human right to information, the environment and information about the environment: From the Universal Declaration to the Aarhus Convention. *Communication Law and Policy, 14*, 73–103.

Day, C. (2015). The Aarhus Convention and NGOs. In C. Banner (Ed.), *The Aarhus Convention: A guide for UK lawyers* (pp. 181–198). Hart.

Dellinger, M. (2012). Ten years of the Aarhus Convention: How procedural democracy is paving the way for substantive change in national and international environmental law. *Colorado Journal of International Environmental Law and Policy, 23*(2), 309–366.

Depoe, S. P., Delicath, J. W., & Elsenbeer, M. A. (Eds.). (2004). *Communication and public participation in environmental decisionmaking*. SUNY Press.

Dobson, A. (2003). *Citizenship and the environment*. Oxford University Press.

Eckersley, R. (2004). *The green state: Rethinking democracy and sovereignty*. MIT Press.

Etemire, U. (2023). Public voices and environmental decisions: The Escazú Agreement in comparative perspective. *Transnational Environmental Law, 12*(1), 175–199.

Falkner, R. (2021). *Environmentalism and global international society*. Cambridge University Press.

Falkner, R., & Buzan, B. (2019). The emergence of environmental stewardship as a primary institution of global international society. *European Journal of International Relations, 25*(1), 131–155.

Falkner, R., & Buzan, B. (Eds.). (2022). *Great powers, climate change, and global environmental responsibilities*. Oxford University Press.

Fasoli, E., & McGlone, A. (2018). The non-compliance mechanism under the Aarhus Convention as 'soft' enforcement of international environmental law: Not so soft after all! *Netherlands International Law Review, 65*, 27–53.

Getliffe, K. (2002). Proceduralisation and the Aarhus Convention: Does increased participation in the decision-making process lead to more effective EU environmental law? *Environmental Law Review, 4*, 101–116.

Hedemann-Robinson, M. (2022, June). Access to environmental justice and European union institutional compliance with the Aarhus Convention: A rather longer and more winding road than anticipated. *European Energy and Environmental Law Review*, 175–188.

Hurrell, A. (2007). *On global order*. Oxford University Press.

Jackson, R. (2003). *The Global Covenant: Human Conduct in a World of States*. Oxford University Press.

James, A. (1986). *Sovereign statehood: The basis of international society*. Allen and Unwin.

Jendrośka, J. (2009). Public participation in the preparation of plans and programs: Some reflections on the scope of obligations under Article 7 of the Aarhus Convention. *Journal for European Environmental and Planning Law, 6*(4), 495–515.

Jendrośka, J. (2012). Citizens' rights in European environmental law: Stocktaking of key challenges and current developments in relation to public access to information, participation and access to justice. *Journal for European Environmental and Planning Law, 9*(1), 71–90.

Kingston, S., Wang, Z., Alblas, E., Callaghan, M., Foulon, J., Daly, C., & Norris, D. (2022). Europe's nature governance revolution: Harnessing the shadow of heterarchy. *International Environmental Agreements*, pre-press online, 1–32.

Lee, M. (2023). The Aarhus Convention 1998 and the environment act 2021: Eroding public participation. *Modern Law Review, 86*(3), 756–784.

Linklater, A. (2011). *The problem of harm in world politics: Theoretical investigations*. Cambridge University Press.

Linklater, A., & Suganami, H. (2006). *The english school of international relations: A contemporary reassessment*. Cambridge University Press.

Navari, C. (Ed.). (2021a). *International society: The English school*. Palgrave.

Navari, C. (2021b). *The international society tradition: From Hugo Grotius to Hedley Bull*. Palgrave.

Okowa, P. (1997). Procedural obligations in international environmental agreements. *British Yearbook of International Law, 67*(1), 275–336.

Pallemaerts, M. (2011a). Introduction. In M. Pallemaerts (Ed.), *The Aarhus Convention at ten: Interactions and tensions between conventional international law and EU environmental law* (pp. 1–16). Europa Law.

Pallemaerts, M. (Ed.). (2011b). *The Aarhus Convention at Ten: Interactions and tensions between conventional international law and EU environmental law*. Europa Law.

Pedersen, O. W. (2008). European environmental human rights and environmental rights: A long time coming? *Georgetown International Environmental Law Review, 21*, 73–111.

Peters, B. (2018). Unpacking the diversity of procedural environmental rights: The European Convention on Human Rights and the Aarhus Convention. *Journal of Environmental Law, 30*, 1–27.

Roach, S. (2005). Value pluralism, liberalism, and the cosmopolitan intent of the International Criminal Court. *Journal of Human Rights, 4*, 475–490.

Ryall, Á. (2023). A brave new world: The Aarhus Convention in tempestuous times. *Journal of Environmental Law, 35*, 161–166.

Samvel, G. (2020). Non-judicial, advisory, yet impactful? The Aarhus Convention compliance committee as a gateway to environmental justice. *Transnational Environmental Law, 9*(2), 211–238.

Shapcott, R. (2008). Anti-cosmopolitanism, pluralism and the cosmopolitan harm principle. *Review of International Studies, 34*, 185–205.

Shelton, D. (2006). Human rights and the environment: What specific environmental rights have been recognised? *Denver Journal of International Law and Policy, 35*(1), 129–171.

Shelton, D. (2010). Developing substantive environmental rights. *Journal of Human Rights and the Environment, 1*(1), 89–120.

Smyth, C. (2022). Marking out the interpretive possibilities of the Aarhus Convention. *Journal of Environmental Law, XX*, 1–10.

United Nations. (2011, July 1). *The Secretary-General: Message to fourth session of the Meeting of the Parties to the Convention on Access to Information, Public Participation in Decision-Making and Access to Justice in Environmental Matters (Aarhus Convention)*, Chisinau. https://unece.org/fileadmin/DAM/env/pp/mop4/Statements/SG_statement_approved.pdf. Accessed 22 May 2023.

United Nations. (2020). *World economic situation and prospects*. https://www.un.org/development/desa/dpad/wp-content/uploads/sites/45/WESP2020_Annex.pdf. Accessed 9 March 2023.

United Nations Economic Commission for Europe. (1998, June 23–25). *Report on the fourth ministerial conference 'Environment for Europe', Aarhus, Denmark*. ECE/CEP/41. https://unece.org/fileadmin/DAM/env/documents/1998/ece/ece.cep.41.e.pdf. Accessed 4 April 2023.

United Nations Economic Commission for Europe. (2023). *Aarhus Convention: Status of ratification*. https://unece.org/environment-policy/public-participation/aarhus-convention/status-ratification. Accessed 9 March 2023.

United Nations General Assembly. (1992). *Report of the United Nations Conference on Environment and Development*. A/CONF.151/26 (Vol. I). https://www.un.org/en/development/desa/population/migration/generalassembly/docs/globalcompact/A_CONF.151_26_Vol.I_Declaration.pdf. Accessed 9 March 2023.

United Nations Treaty Collection. (2023). *Convention on access to information, public participation in decision-making and access to justice in environmental matters: Status of ratification*. https://treaties.un.org/Pages/ViewDetails.aspx?src=IND&mtdsg_no=XXVII-13&chapter=27#1. Accessed 9 March 2023.

Wapner, P. (1996). *Environmental activism and world civic politics*. SUNY.

Warren, M. (2017). *The all-affected interests principle in democratic theory and practice* (Working Paper). Institute for Advanced Studies, Vienna. http://aei.pitt.edu/93142/1/Warren_-_All_Affected_Interests_Principle.pdf. Accessed 8 May 2023.

Wates, J. (2005). The Aarhus Convention: A driving force for environmental democracy. *Journal of European Environmental and Planning Law, 1*, 2–11.

Weaver, D. (2018). The Aarhus Convention and process cosmopolitanism. *International Environmental Agreements, 18*(2), 199–213.

Wheeler, N. (1996). Guardian angel or global gangster: A review of the ethical claims of international society. *Political Studies, XLIV*, 123–135.

Wheeler, N. (2000). *Saving strangers: Humanitarian intervention in international society*. Oxford University Press.

Wheeler, N., & Dunne, T. (1996). Hedley Bull's pluralism of the intellect and solidarism of the will. *International Affairs, 72*(1), 91–107.

Weinert, M. S. (2011). Reframing the pluralist-solidarist debate. *Millennium, 40*(1), 21–41.

Wight, M. (1991). *International theory: The three traditions*. Leicester University Press.

CHAPTER 2

Theoretical Framework

Introduction

In the preceding chapter (see Chapter 1), I introduced Aarhus (Aarhus Convention, 1998) as a potentially pioneering agreement. I argued that it propagated "humankind" concerns for empowerment, "voice" and co-creation, in a way that avoided attenuating the primacy of state sovereignty. I argued that sovereignty—the legitimacy and authority with which states govern *domestically* and are recognised as legitimate actors *internationally*—concomitantly (a) undergoes ethical maturation and (b) regulates the extent to which such ethical ambitions grow. As such, I implied interplay between solidarist ambition and pluralist reality. I argued that both regulate one another. This was cast as consistent with the logic of solidarisation.

These ideas—of pluralism, solidarism and solidarisation are associated with the English School of IR. The English School traces its roots to the British Committee on the Theory of International Politics (Dunne, 1998; Epp, 2017). Whilst it enjoys contemporary popularity far beyond the borders of the UK (see Flockhart & Paikin, 2023), the English School historically had a notable presence at London School of Economics, University of Oxford, and the Department of International Politics, Aberystwyth University. I regard as a chief premise of English School thinking the conviction that whilst anarchy (the absence of world government) cannot be avoided, the risks of anarchy can be alleviated, to some

D. Weaver, *The Aarhus Convention*, Environmental Politics and Theory, https://doi.org/10.1007/978-3-031-43536-2_2

extent, by a commitment by states to some greater or lesser degree of convergence on issues of shared concern (see as an initial starter Bull, 2002; Buzan, 2004, 2014; Hurrell, 2007; Linklater & Suganami, 2006). When states participate in such convergence, they form an international society. It is in this context that the English School is sometimes termed the international society approach (Navari, 2021a, 2021b).

Institutions tend to be deemed by English School thinkers to hold normative worth (Brems Knudsen, 2019; Suganami, 1983). Primary institutions like state sovereignty hold together, and sustain the existence of, IR as we know it. I deem sovereignty the definitive currency of IR, invested by states in pursuit of their interests. Indeed, a key theme ebbing throughout this book is the contingency of ethical progress upon state consent. The voluntarism associated with sovereignty will surface on a number of occasions as the book gains traction: states will be seen to withdraw their consent from multilateralism, at a number of points, when they do not deem multilateralism to serve their interests.

Secondary institutions, such as international environmental agreements, operationalise primary institutions; secondary institutions offer the "tools" that "get IR done" (see Ahrens, 2019; Brems Knudsen, 2019; Falkner, 2021; Falkner & Buzan, 2019; Hurrell & Kingsbury, 1992). Interdependence exists between primary and secondary institutions. On one hand, secondary institutions would not exist without primary institutions breathing life into them. An international environmental agreement would not exist without consenting states investing their sovereignty in it, in a mutual attempt to address a particular issue. Yet at the same time, secondary institutions can yield a reforming, ethically additive influence on primary institutions (Ahrens, 2017, 2018, 2019; Ahrens & Diez, 2015; Brems Knudsen, 2019; Falkner, 2021).

This latter point is prescient for solidarisation, explored later. To reach that discussion, it is necessary to elucidate the English School continuum of realism, rationalism and revolutionism (Wight, 1991). The continuum offers three competing, complementary imaginaries of IR, with which to assess "real world" developments. They do not exist in a zero-sum game where one triumphs over the others. In fact, I personally avoid speaking of "international society", "international system" or "world society" as existing definitively and encapsulating "one world". I find it more accurate to speak of international society *arrangements*, existing in multiple, diverse configurations simultaneously. The reality of IR, after all, is that multiple system/society/cosmopolis arrangements coexist, given the inherent pluralism and diversity of the world's two-hundred-odd states and myriad human collectivities.

To demonstrate this, it should be clear that rationalist legal and diplomatic arrangements coexist with realist arrangements of war. Warfighting and peace-making ensue concomitantly. In *some* circumstances, at *some* times, *some* states struggle against others for their existence, given the absence of a world government to curtail their enmity. These are realist arrangements (Mearsheimer, 1990; Morgenthau, 1985; Waltz, 1979). But whilst such enmity ensues, attempts to beat swords into ploughshares continue, despite difficulty (Claude, 1984). The invaluable work of negotiating *agreement*, if not *peace*, between states does not cease because of war: if anything, it intensifies. And at the very least, it continues in the hope of identifying issues of shared concern, upon which states can converge in international law and international organisation (for the latter, see Armstrong et al., 2004; Roberts & Kingsbury, 1993).

In the midst of international conflict and convergence, even the faintest glimmers of revolutionist cosmopolis appear. World society arrangements coexist with international system and society arrangements. We may consider, in this regard, the most radical elements of the transnational environmental social movement demanding state-societal "system change" (Weaver, 2022). It is possible to witness the prefiguration of alternative orders by nonstate actors who, for instance, imagine (and attempt to implement) alternative currencies, decision-making practices and governance modalities, with a view to usurping sovereignty (Berglund & Schmidt, 2020; Wall, 2002; Weaver, 2022). Let me reaffirm the point: realism, rationalism and revolutionism are not "either/or" binaries. Evidence of each emerges concomitantly in the diverse, heterogeneous "real world". The English School analyst's task is to seek evidence of them, and to gauge the extent of their relative potentialities in the object under investigation.

Wight (1991: 260) would not, I hope, be entirely opposed to my understanding of the continuum, given his position that realism, rationalism and revolutionism are like "streams…interlacing and never for long confined to their own river bed". That said, most English School readers will agree that the English School was long associated with a preference for international society arrangements, and continues to be equated, in the introductory seminar at least, with the "international society approach" (Navari, 2021a, 2021b). International society, a reasonable bridge between realism and revolutionism, imagines arrangements in which, "despite the formally anarchical structure of world politics, interstate relations are governed by normative principles in light of which states can, and to a remarkable degree do, behave reasonably towards one another" (Linklater & Suganami, 2006: 29–30). This is persuasive: to heed the possibility of such "reasonable" arrangements is to acknowledge the inevitability of anarchy, whilst investing cautious hope in the possibility that states are willing and able to converge, to some extent. I regard the logic of international society as realist*ic*: it accepts the lasting predominance of sovereign statehood (James, 1986) whilst recognising that states, sometimes in concert with nonstate actors, can and *do* achieve some limited modicum of ethical progress. Neither is this understanding naïvely utopian nor unduly dystopian.

With this in mind, the chapter will pay attention to the two understandings of international society that emerge from the English School. Pluralism, leaning towards the realist end of the continuum, imagines a world of states' minimal coexistence, "satisficing" their obligations to preserve themselves and their interests. Pluralist arrangements seek to reduce the risk of international conflict and preserve IR's status quo, in a bid to avert international disorder. Pluralism is state-oriented, hinging as it does on voluntarism: the willing participation of states in multilateralism, in the knowledge that they can withdraw at any point with impunity.

Solidarism, meanwhile, leans towards the revolutionist end, whilst remaining firmly located within rationalism's boundaries. My understanding of, and preference for, solidarism is firmly state-oriented, but more ethically ambitious and demanding of states. It demands a responsible sovereignty, leveraging it for more ambitious other-oriented interests. Good international citizenship plays a key role here. To be a good international citizen is to balance pursuit of one's national interest, international society's prevailing norms, and "cosmopolitan ethical responsibilities to the wider human community" (Gilmore, 2023: 84). If statist self-concern animates pluralism, more ambitious other-oriented concern animates solidarism. I must stress that whilst revolutionism attempts to unshackle itself from sovereignty, state-oriented solidarism remains anchored to sovereignty, avoids usurping it, and enriches statist IR without "undoing" it. This is my understanding of, and normative disposition towards, state-centric solidarism, employed in this book. It is an approach to international society that inclines states towards other-oriented concerns, as opposed to solely state-oriented concerns.

What remains, in this chapter, is an assessment of each element of the continuum. Both extremes, realism and revolutionism, are assessed first. Rationalist international society is examined afterwards. There, discussion of solidarisation ensues. I find solidarisation to be compelling, in its presupposition of interdependence between pluralism and solidarism. In the *process* of solidarisation, solidarist progress is propagated inside the parameters of pluralism. Co-constitution between pluralism and solidarism is evident, given the viability of pluralist arrangements (a) being rendered more ethically mature whilst (b) containing and confining solidarist ambition, in order to avoid the undermining of sovereignty and the resultant destabilisation of IR.

Realism

Realism is a conflictual, statist understanding of IR that foregrounds the autonomous, atomised state seeking survival in a world of anarchy, the absence of world government. A central focus is on this absence of supranational authority, and the resultant impossibility of curtailing international harm between states. It is posited that whilst states can yield control domestically, no such control over states occurs supranationally. There is no global Leviathan. For this reason, associations between realism and Hobbesian thought were critically engaged by English School thinkers (Bull, 1981; Vincent, 1981). The corollary of anarchy is states' necessity to pursue, as a chief priority, their national interests. A state's capacity and willingness to use force to protect its territory, population and interests is of key importance, under realism. States form alliances and superpower blocs as the balance of power recalibrates according to changes in, or changes to the equivalence of, states' power and force endowments (Gaddis, 1987; Mearsheimer, 1990; Rosecrance, 1966; Waltz, 1979). But ultimately, states are "on their own" in realist arrangements. Realism thus does not invest undue hope in states converging with one another. It has been said, after all, that anarchy compels them to be "death-fearing and power-seeking" (Laferrière & Stoett, 1999: 84).

Such an ontology of survival and conflict presupposes the sanctification and protection of states' territorial integrity. This cannot be achieved without clearly delineated, fortified borders. Bordering processes delineate inside/outside, domestic/foreign, home/away, and are integral to states' articulation of the friend/enemy distinction (Schmitt, 2005, 2007). It is, as such, the case that realism's geopolitical concerns had nationalist connotations, given the foregrounding of domestic, territorially aligned selfhood, literally and figuratively fenced against the external domain "beyond the border". Such delineation of the world into a "grid" of atomised states is Westphalian in rendering the world one of "absolute spaces and explicit, non-overlapping boundaries", of "moral geographies of similarity and difference, inclusion and exclusion, which sharply distinguished 'us' from 'them'" (Warf, 2009: 64). The early *Geopolitik* (Ancel, 1936; Curzon, 1907; Hartshorne, 1939; Haushofer, 1932; Holdich, 1916; Kjellen, 1915; Mackinder, 1904; Ratzel, 1897; Schmitt, 2003, 2005, 2007) thus yielded environmental connotations, but it needs very sensitive treatment given its historical contexts. Weight was placed, there, on preservation of sovereign territory, national cohesion and territorial gain. Such fetishisation of sovereignty and territory precludes pursuit of anything much more than survival, perpetuating a system of fixity (Weaver, 2020). This is far from an optimistic understanding of the human condition, and rather one driven by the *animus dominandi*, the human lust for power (Morgenthau, 1945).

From the above, we might suppose that the discords between realism and environment drone out the harmonies. But if we start from the position that states engaged in realist arrangements need to protect *territory*—their earth, air, water—the harmonies begin to emerge. Territorial integrity is environmental integrity. Under this logic, the borders of state territory, of sanctified national environment, are "power containers" (Agnew & Corbridge, 1995: 93–94; Giddens, 1987: 49; Taylor, 1994).

It is also the case that the national interest has environmental qualities. Environmental change and protection, resource scarcity and ecological harm have figured in the environmental security literature for decades (see amongst many others Barnett, 2003; Dalby, 2000; Gleick, 1991; Homer-Dixon, 1994, 1999; Kaplan, 1994; Myers, 1993). Security actors—not least states' military and intelligence apparatuses—assigned environmental protection as a priority, and allocated appropriate resources to address it, long ago. Realists would argue that such environmental securitisation (see Buzan et al., 1998) is impossible without the national security prerogatives, urgency, and *Macht*-mindset of realism. They would also stress the necessity of a hawkish focus on statecraft and strategy, power balancing, and alliance formation, if states are to effectively mobilise their security resources for environmental protection. They would deem realism a prudent, power-minded approach to "getting things done" with urgency and, when necessary, force.

However, realism is not immune to critique by environmentalists. Firstly, realism's war-readiness is not aligned with the environmental preference for saving non/human life. War is destructive. Secondly, critics may observe realism's implication in marginalisation: its war-readiness places heavier weight on men, given their traditional warfighting primacy. Women, until recently, were marginalised and excluded from frontline fighting, although times are changing (Clark, 2022). There is a case nonetheless that the rhetoric of realism—survival, strength, force—privileges robust masculinity and marginalises that which is not congruent therewith. Thirdly, *epistemological* discords separate realism and environmentalism. Realists often avoid normative assessments of "what ought", preferring objective treatment of "what is". There is a realist preference for establishing "the facts" of IR, accompanied by a perennial lesson that "history repeats itself" (Koliopoulos, 2019). Realism lacks cognitive space for radical alternatives: for imagining what *could be*; for seeking the "oughts" of world affairs and ways to achieve them. Realism is susceptible to charges of defeatism in its resignation to a world of conflict between bordered states. Environmental harm knows no borders: it cares not for traditional delineations of inside/outside, self/other (Hurrell, 2007). Global harms like climate change and extinction have not the slightest congruence with artificially imposed borders. Equally, those human actors seeking to address such harms neither tend to be bound by such borders in an age of globalisation (Archibugi & Held, 1995; Castells, 1996, 1997). Humans and their associations are global actors, struggling to achieve global purposes, coordinating themselves above, below, through and beyond borders. States' ability to control them has arguably been in decline for some decades. Bearing this in mind, the chapter now turns to the other end of the continuum, revolutionism: an imaginary that foregrounds the emancipation of humankind from sovereignty.

Revolutionism

Those occupying the revolutionist end of the continuum will be unconvinced and disquieted by the realist imaginary. Their critique will firstly relate to realists' preoccupation with sovereignty and territory. Borders, revolutionists would argue, are not fixed; neither should they be. Humankind—alongside the biosphere—came long before the artificial impositions of states and borders. Realism, when subjected to such critique, falls into and perpetuates the "territorial trap" (Agnew, 1994;

Agnew & Corbridge, 1995: 83), which casts sovereign territory as *the* definitive ontological starting point of IR: sacred and immutable. By falling (and dragging analysts) into this trap, realism would be seen to essentialise IR, rendering unchangeable the domestic/alien binary delineation.

To fetishise sovereign territory is to ignore the ceaseless, unbounded flows of power, progress and interaction between humans across the world. Sovereignty, national identity and geopolitical boundedness (see Paasi, 1999, 2001, 2009) are usurped by revolutionist arrangements. The goal is a utopian endgame, cosmopolis: a community of humankind that has unshackled itself from sovereignty. I reaffirm, here, that revolutionism is different from my understanding of state-oriented solidarism, because solidarism remains firmly bounded by sovereign architectures, and does not seek usurpation, but rather incremental reform within those architectures. Whilst revolutionism is oriented around *emancipation*, my understanding of solidarism is concerned for *empowerment*, which eclipses the most scant indications of emancipation observed in the Aarhus Convention. Rather than imagining a rationalist international society prevailed upon by states, revolutionism analytically foregrounds a world society—in my mind, cosmopolis—that is constituted by "non-state actors such as NGOs...individual human beings and the idea of a global human community" (Williams, 2013: 127). Such a world society embraces "the totality of global social interaction" (Bull in Williams, 2013: 130) and thus foregrounds humans as opposed to states. Revolutionism is therefore post-sovereign in its inclination towards a Kantian form of universal human community that rids itself of its chains.

I must reassert, here, that whilst my present focus is on the empowerment of humans and their associations—world society actors, if you will—any such empowerment is limited within distinctly statist international society arrangements, delineated by Aarhus' Party jurisdictions. My present study does not observe an emancipation of world society actors 'breaking free' from sovereignty. The book observes a *cosmopolitan impulse* driving solidarist ethical progress within international society arrangements, rather than the above form of utopian, post-sovereign endgame.

Bearing in mind such utopian post-sovereign thoughts, it is possible to identify harmonies between revolutionism and the Aberystwyth school of critical IR and security (Booth, 1991, 1997, 2007; Krause & Williams, 1997; Linklater, 1998, 2002, 2009). Ken Booth himself (1997: 110) defined emancipation as "freeing people, as individuals and groups, from the social, physical, economic, political and other constraints that stop them from carrying out what they would [otherwise] freely choose to do". Such an ethic of human care would apply to distant, unknown, and yet unborn strangers as much as to co-nationals (see Shapcott, 2008: 196). In order for such emancipatory concerns to be universalised, three tenets are required, argues Pogge (1992): first, the referent of concern must be the human individual; second, there must be total equality in the attachment of this status to all humans; and third, such status must have global force. Under this logic, one humankind occupies one planet, the singular Spaceship Earth (Selcer, 2018), meriting a cosmopolitan ethic that avoids consideration of exactly *who* the individual human referent of concern is. *Every* occupant aboard Spaceship Earth is of equal vulnerability and worth. This echoes Linklater's (2007: 138) finding that "mutual recognition of shared… vulnerability provides the most readily available means of projecting forms of solidarity across the boundaries of established communities".

If such emancipation—such 'freedom from' and 'freedom to'—is pursued with total zeal, the state materialises as potential hindrance rather than help. Earlier, I referred to the prefiguration of post-sovereign alternatives: the 'living out' of radical futures. This was associated with the radical environmental movement (Weaver, 2022). There are indeed links between emancipation and environment. Linklater (2006: 110) regarded global environmental harm as stimulating "communities of fate or risk which are more inclusive than any national political order; they have promoted the globalisation of shared sentiments". This was said to evoke "cosmopolitan identities which transcend the parochialism of the nation state" (Linklater, 2006: 118). This gels with Beck's (1999: 62) claim that with global environmental harm comes 'the end of the Other', and surely so out of necessity: were the planet to face true existential crisis, complete human unity would be required. This would nullify the value of those artificially constructed b/orders mentioned above. Such artificial impositions, in circumstances of unequivocal global vulnerability, would constitute impediments to necessarily global remedial interventions.

Revolutionism thus has implications for both the global and local scales. Globally, it would entail envisaging a utopian cosmopolis, a 'humanity beyond borders', that is emancipated from sovereignty. Figuratively, this achieves limited traction in IR given the power, for instance, of social media, conventional media, the internet and telecommunications, and the related capacities for human agency to be propelled globally without impediment by borders. Yet literally, I doubt that any such global utopian progress has been achieved, or is possible. The residual persistence of sovereignty is such that the world's political arrangements continue to be underpinned by states. Locally, revolutionism has a little more practical purchase, although significant contemporary subversions of sovereignty will not survive to any truly durable extent, if we are using the above notion of utopian human 'freeing'. Local revolutionism may materialise in a patchwork of prefiguration, in small scale experiments that seek to imagine and operationalise the relinquishment of sovereignty.

Assistance in gaining theoretical traction, in this regard, is found in Falkner's (2021: 63) understandings of eco-localism and eco-globalism. Eco-localism would be characterised by heterogeneous grassroots practices, with local communities forging their own environmental realities, whilst eco-globalism would involve a universal human community pursuing environmental values in pursuit of "a shared vision of global environmental sustainability" (Falkner, 2021: 63). Both such imaginaries gel with the revolutionist end of the continuum. Both involve humans mobilising beyond the strictures of sovereignty for environmental purposes. And both involve humans 'rising' from the grassroots to forge their collective environmental futures, without a state monopoly of governance.

It goes without saying, of course, that revolutionism itself is susceptible to critique. The first relates to the risks of emancipation in a world of market and political liberalism. Human 'freeing' may not beckon green futures. It would be naïve to assume buoyant uptake of green-minded cosmopolitan citizenship (Dobson, 2003), of nascent virtuous environmental citizenries (Connelly, 2006), or of 'biocracy' (Ball, 2006: 145). Laypersons, lacking environmental education and awareness, who struggle to pay their rent, will (continue to) reward themselves with instant material gratification, if they land themselves on some financial good fortune. Such material gratification will (continue to) be high carbon and throwaway. Critics would warn emancipatory thinkers to be 'careful what they wish for', because human 'freeing' might backfire, and do more damage than good. A counterargument, namely that a green revolutionism would insist on *environmentally benign emancipation*, is replete with its own potentially attendant injustices. What of developing countries' pursuit of the same 'coal-fired' growth that advanced economies enjoyed? What of the once impoverished layperson who deems it their natural right to conspicuously consume after meeting wealth? What of the risk that *diverse* and also *non-environmental* voices are silenced, in public discourse, by the tyranny of a dogmatically ecological utopia?

Notwithstanding the difficulties of educating children and adults for sustainability (see Huckle & Sterling, 1996 for a seminal introduction) and weaning them off throwaway material gratification, environmental cosmopolitan citizenship is a utopia that will either fail (a) once practically deployed or (b) before experimentation. This is because the fostering of sincere ethical concern for distant, unknown, even unborn others is prohibitively difficult. It is tough enough, under market liberalism, for next-door neighbours in Commuterville to nod across the garden fence and help each other in emergencies. This difficulty is compounded inordinately when one attempts to *feel for* human life in the abstract.

This leads to the second critique: that the notion of a nascent legion of humankind unshackling itself from sovereignty is not possible. Whilst the major revolution of recent history, October 1917, drove the Soviet Union's (USSR's) emergence, the latter (a) remained totally embedded in the architectures of sovereignty, despite emancipatory visions (Marx & Engels, 2002); (b) failed to globalise; and (c) failed outright, not least given the human and environmental harms of inefficient, environmentally destructive, command-and-control economics and autocracy (see Harman-Stokes, 1995; Honneland, 2003; Ziegler, 1987).

On this point, the third critique of revolutionism is a warning: revolutionist usurpations are dangerous. They are violent and bloody. 'System change', as it is termed nowadays (Weaver, 2022), does not come without cost. Those costs are shouldered by the humans pursuing it. Those orchestrating system change will not always be the ones suffering as a result: they will more likely be political elites, somewhat more immune to the physical risks. In representative democracies, the state is responsible for aggregating and pursuing citizens' interests (Hague & Harrop, 2007; Held, 1993; Lijphart, 1999). Revolutionary actors tend not to have such democratic mandate.

To usurp sovereignty is to abandon the one constant in IR. It is to rescind the good done by states, despite their flaws. In fact, those flaws are less inherent to the analytical unit of state, and more intrinsic to the *very humans* occupying a state's governmental executive. Is it not remarkably risky to relinquish sovereignty altogether, and throw all of one's hope in the emancipatory project? As the next chapter (see this book, Chapter 3) demonstrates, the pursuit of stable international order has underpinned states' multilateral progress. Amidst war, states continue to collectively invest their sovereignty to address common concerns. This is to be celebrated. Without such a voluntarily attempted stabilisation of international order, there would be no Charter of the United Nations (1945), no Universal Declaration of Human Rights (1948), no Paris Treaty Establishing the European Coal and Steel Community (1951), the latter an attempt to curtail the tools of war between European states, and hence reduce the likelihood of human suffering. As is so clearly evident at the time of writing, states' capacity to hurt is vast. But so too is their capacity to do good. To abandon states is to undo centuries' efforts and expose the world to the frightening unknown.

Thus far in the chapter, I sketched the continuum's extremes: a dystopian realism of states in anarchy, and a utopian revolutionism of humans escaping sovereignty. It will come as no surprise that my substantive analysis of Aarhus, which follows this chapter, only identifies very limited indications of realism and revolutionism. Realist concerns are peppered sparingly. We might turn to Belarus' withdrawal from Aarhus (Press Service of the President of the Republic of Belarus, 2022), understood in the context of Belarus ostracising itself from human rights compliance amidst its implication in the Russian Federation's illegal invasion of Ukraine. We might turn to the Russian Federation's refusal to sign Aarhus given its concerns over environmental transparency, accountability and the exposure by environmental defenders of the risks posed by some of its ailing naval assets (Goldman Environmental Foundation, 2023). But by no means do I perceive much of the realist survival impulse, in the analysis of Aarhus that follows.

Neither does the analysis perceive revolutionism. A key point, in this regard, is that environmental civil society is "built in" to the very fabric of environmental multilateralism that constitutes Aarhus' architecture. From Chapter 3 onwards, we will witness the formal inclusion of civil society in the Convention's negotiations (see Chapter 3), procedural rights provisions (see Chapter 4) and organisational infrastructure (see Chapter 5). Humans and their associations have a well amplified voice in the Convention, but it is articulated through existing channels of power. Humans and their associations far from usurp sovereignty: rather, they reinforce it by contributing an ethical additive to the extant machineries of sovereignty. And we will see, in the following chapters, that any such ethical enrichment by human actors is contingent on state consent: the willing participation of states in an international order "within which no sovereign would submit except to those rules to which it had consented" (Hurrell, 2007: 49).

As such, my ethical reading of Aarhus sits most comfortably with the "middle way" of the English School continuum, rationalist international society. Evident, in the Convention, is a counterbalancing between (a) pluralist attempts to preserve states' diverse interests, and to reduce harm sustained by states and (b) solidarist attempts to codify stricter human rights provisions and more justiciably address *human* harm and injustice. This latter approach remains firmly within the bounds of sovereignty. Whilst it might be animated by a cosmopolitan impulse, it leverages human empowerment rather than emancipation within statist arrangements. Let us turn to international society without further ado.

Rationalism

Having discussed the two ends of the continuum, my task now is to assess rationalist international society, long associated with the English School's preference for cautiously exploring ethical potentialities in a persistently state-dominated and anarchical world (Navari, 2021a, 2021b; Wight, 1991: 30–48). International society offers a prudent "middle way" between realism and revolutionism. It seeks compromise between realist survival and human cosmopolis. I follow Bull (2002: 9–10) in discerning international system from society: systems emerge where states' interaction is "sufficient to make the behaviour of each a necessary element in the calculations of the other". Adversarial arrangements require states to monitor, calculate and adapt to one another's capabilities and signals. Such posturing and balancing is essential for survival and the achievement of national interests. However, the logic of international society, whilst accepting anarchy, does not resign itself to the inevitability of calculation and conflict. It cautiously accepts the possibility of states *rationalising* their affairs with one another, and seeking some modicum of progress. Limited space is opened to consider how states may—to some lesser pluralist or greater solidarist degree—converge and form *societies*, in pursuit of common ends. International societies are characterised by states' mutual commitment to norms "that both constitute and regulate the relationships" binding them (Williams, 2013: 129).

I find reasonableness and pragmatism in this. Such logic is realist*ic*: neither resigning itself to survivalist dystopia nor seeking the utopian revolutionist unknown. To subscribe to such rationalist, prudent logic is to heed that "despite the formally anarchical structure of world politics, interstate relations are governed by normative principles in the light of which states *can*, and to a remarkable degree *do*, behave reasonably towards one another" (Linklater & Suganami, 2006: 29–30; emphasis added). With this in mind, recall Bull's (2002: 13) observation that international societies emerge when a group of states, "conscious of certain common interests and common values, form a society in the sense that they conceive themselves to be bound by a common set of rules in their relations with one another, and share in the working of common institutions".

Central to international society is the norm of sovereignty (Jackson, 2003; Williams, 2013). When states deliberately form international societies, they accept one another as sovereign, in terms of their domestic authority to govern and their international legitimacy as autonomous actors. They "invest" their sovereignty collectively, to some greater or lesser extent, in order to address shared concerns. When forming such societies, states acknowledge the risks of anarchy but consciously attempt to mitigate them, by leveraging their sovereignty together. It is for this reason that I earlier defined an international society as a deliberate convergence of states upon issues of shared concern, involving states' conscious multilateral investment of sovereignty in measures to address those issues. Shared norms are placed on collective agendas; efforts are taken to codify them using the tools of international law and diplomacy. In this attempt at rationalising international affairs, sovereignty remains the currency, but is spent for *inter*national as well as national interests. Evident here is a cautious hope, grounded by a certain doubt about the feasibility and appropriateness of meddling too much with the political world's statist structures. Sovereignty manifests itself as a primary institution, without which secondary institutions—international environmental agreements, in our case—would lack states' "buy in". Buzan (2014: 16–17) casts primary institutions, such as sovereignty, as "deep and relatively durable social practices in the sense of being evolved more than designed. These practices must not only be shared amongst the members of international society, but also be seen amongst them as legitimate behaviour". Secondary institutions, such as international environmental agreements, are tools deliberately intended to help states regulate their convergence (Falkner & Buzan, 2019: 134).

Having cleared some initial ground, it is now necessary to clarify the pluralist and solidarist understandings of international society for which the English School gained fame (Weinert, 2011). From the outset, let me clarify pluralism as an approach that foregrounds coexistence between states seeking to preserve their diversity, and hence their self-identities and self-interests. The focus, in pluralism, is on preventing harm caused and sustained by states, in order for them to enjoy as stable an international order as possible. Pluralism sits towards the realist end, given the centrality of states' self-orientation and preservation of their own prerogatives. But it remains firmly embedded in rationalist international society. After all, states engaged in pluralist arrangements seek to overcome anarchy in their deliberate convergence.

My understanding of solidarism, meanwhile, is positioned towards the revolutionist end, but again is lodged firmly in the parameters of international society. Solidarism seeks to propagate norms that go beyond states' self-interest, challenging states to develop other-orientations and (collectively) invest their ethical attention in *human* as opposed to solely state vulnerability. My understanding of such "humankind" solidarism, retaining sovereignty but ethically agitating it, brings to the fore potential internationalisations of human rights. It allows space for pursuing international justice values as opposed to only international order (Linklater & Suganami, 2006: 199–206; see Bain, 2021). What follows elucidates this distinction further.

Pluralism

The pluralist imaginary is "thinner" than solidarism, given that states engaged in pluralist arrangements share fewer values, and that their chief objective is to achieve mutual coexistence (Buzan, 2004: 59). More optimistically than realism, pluralism invests modest hope in possibilities for states limiting the harm they cause each other. Pluralism was thus associated with *international* harm conventions, as opposed to solidarism being associated with *cosmopolitan* harm conventions (Linklater & Suganami, 2006: 8). This pluralist concern for international harm places gravity on non-intervention and preservation of states' territorial integrity (Ahrens & Diez, 2015: 343). If my understanding of solidarism shares affinities with, and retains ethical space for, moral cosmopolitanism within statist parameters, pluralism is more communitarian in outlook, favouring the diversity of bounded, autonomous states, seeking to preserve their heterogeneities and hence their pursuit of particular interests (Brown, 1992; Rengger, 1996). Pluralists make the case that whilst group affiliation ascribes meaning to humans' lives socially and politically, attenuating some of the differences between group members, it also "sharpens the distinctions and deepens the divisions that separate people whose lives are located on different sides of the lines that are drawn" (Jackson, 2003: 402).

With this in mind, pluralism evinces a "live and let live" ethic wherein states seek to preserve their diversity, iterating their own prerogatives over more ambitious other-oriented ones. It imagines a world wherein "each group can sit undisturbed under their own fig tree" (Hurrell, 2007: 299). Or, to stay for a moment in the garden, it espouses a "morality of 'tending your own patch'...that means having a patch and being free to occupy it and cultivate it in your own way" (Jackson, 2003: 410). Pluralists doubt the feasibility and appeal of imposing universal norms on particular jurisdictions. They regard the impression of one 'good life' onto diverse states as artificial and injurious, given the risk that heterogeneity would be silenced. To tarnish the rich and diverse world with a 'one size fits all' ethical brushstroke would be to homogenise, essentialise, and render artificially uniform human relations and IR. Bain (2021: 100; emphasis added) captures this by observing that states, for pluralists, each host "a *particular* conception of the good life". Such is the communitarian recognition that boundedness and delimitation contribute to the persistence of diverse national identities and the territorial containment of citizens' and states' ethical concerns (Dobson, 2006: 222–223; Eckersley, 2006: 93). This is not necessarily negative: pluralists are disposed to appreciate the "unavoidable realities of human diversity and human imperfection" and of how they necessitate a pragmatic "workable ethics" in IR (Jackson, 2003: 400).

Such 'live and let live' ethics require foregrounding sovereignty, in order to maintain states' mutual respect for territorial integrity and non-intervention. The corollary is the importance of states' willing consent and voluntarism: their ability to enter and withdraw from the diplomatic 'pitch' at will. It is crucial for multilateral progress, under pluralism, not to go 'too far' in agitating states' ethical thresholds. Any ethical focus will ultimately centre on avoiding destabilisation and resultant harm to states. States constitute the chief referent of ethical concern. International interests are formulated to serve states' *national* priorities. States voluntarily follow international interests, under pluralism, to the extent that it is "*our* [state-specific, particular] creation and acceptance of the norms of international society that alone creates the obligation to follow them" (Rengger, 1996: 72; emphasis added).

Environmental implications can be elicited from the above. International efforts to protect the environment remain state responsibilities. Creation of international environmental agreements, and associated use of diplomacy and international law, falls within states' purview. Sovereignty is consciously invested by states, in order to promote and protect their interests, economies and citizenries from environmental decline. Such deliberate, if limited, investment of sovereignty is consistent with what Falkner (2021: 63) termed "Green Westphalia". We would do well to follow Falkner's (2021: 64) finding that states, when mobilising on environmental issues, evince environmental *raison d'état*. In pluralism, states seek to avoid harming *one another* environmentally, working to reduce and prevent transboundary harms whilst seeking to retain their autonomy and control inside their borders (Falkner, 2021: 66). Pluralist arrangements would also evince an environmental *raison de système* in which states cooperate minimally to preserve the environmental health of their international society arrangements, mitigating the environmental risks posed to the world and its sovereign constituents (Falkner, 2021: 229). The pluralist ethical driver is the prevention of international disorder resulting from environmental harm.

A green pluralism would thus seek to sustain international order and prevent *dis*order arising from environmental harm. States act 'beyond the border' not to altruistically repent for the harm they caused the non/human environment, but because it is in their interests to do so. Stable international order necessitates coexistence between, and convergence by, states on environmental matters. Environmental health manifests itself as a prerequisite for preserving sovereignty. Sovereignty cannot be invested if environmental risks jeopardise either the structures of international society, or states' functional continuity. And neither can the natural environment be managed and protected to its fullest extent unless a sufficiently resilient and robust sovereignty is preserved and leveraged accordingly. Environment and sovereignty materialise as interdependent. In this regard, I join Falkner (2012: 516) in foregrounding "sovereignty as a defining principle of environmental management". But under pluralism, any international environmental cooperation needs to be

(a) inherently palatable for states and (b) sufficiently accommodating of their diversity, allowing enough leeway between states to leave "room for the tribes" (Walzer, 1994: 64).

Solidarism and Solidarisation

If the prevention of ***international*** harm between states is a pluralist cornerstone, the prevention of ***human*** harm underpins my understanding of solidarism (Linklater & Suganami, 2006: 8). Whilst less intrusive solidarist approaches may simply demand more extensive cooperation between states, my more demanding approach retains enough ethical space to propagate moral cosmopolitan values within the boundedness and structural limits of sovereignty. Harm reduction surpasses territorial non-intervention and integrity. It involves generating ethical concern for humans beyond as well as within one's borders (Ahrens & Diez, 2015: 343).

Presciently for Aarhus, such solidarism casts humans themselves as subjects of international law, necessitating realignments of sovereignty (Ahrens & Diez, 2015: 343–344). A mechanism for this is the codification of the human ability to claim rights vis-à-vis the state (Buzan, 2004: 149; Donnelly, 1998: 21; Wheeler, 2000). Pluralists may deem this destabilising, as such rights-*claiming* requires (a) the practical use of legal instruments that might attenuate states' diversity and (b) the willingness to impose universal norms on particular and heterogeneous jurisdictions. Pluralists would, furthermore, be sceptical of the solidarist push towards "human perfectibility and progress", given that it "soars above human history and experience, above reality" (Jackson, 2003: 401).

Yet solidarists would refute this, I think, by arguing that the benefits of codifying such provisions, which at least aim to protect all human life, outweigh the costs. They would turn to the reality, if not success, of such tools as the Universal Declaration of Human Rights (1948), humanitarian intervention (Wheeler, 2000) and Responsibility to Protect (R2P) doctrine (ICISS, 2001). Such tools seek to preserve the worth of human life, implicating all states in the protection thereof, irrespective of states' diversities. By declaring, for instance, their commitment to the Universal Declaration, states committed (if but formally) to recognising the importance of not deliberately and unlawfully harming innocent life and jeopardising the rights they declared "fundamental". Likewise, R2P obliges states to act—if necessary, with force—when the most egregious human rights violations occur. This is emblematic of state-centric solidarism: sovereignty remains strong enough to discharge itself forcefully, but is harnessed to *protect* rather than *take* life. Sovereignty materialises as a "force for good". It is no coincidence that Ahrens and Diez (2015: 345) deemed sovereignty and human rights, under solidarism, "mutually constitutive". The logic is that states must "work more" to retain their legitimacy, when participating in solidarist arrangements. A more responsible sovereignty, under solidarism, manifests itself not as an absolute given, but as conditional on avoiding harm to innocent life.

As the above shows, and importantly for Aarhus, such provision of human rights is not contained by or within borders. It goes 'beyond the border', avoiding national/foreign distinctions. The human is no more or less worthy of rights-claiming because s/he is a citizen of a particular state, under solidarism. This dovetails with a recognition that global harms are, by definition, immune to border fortification. 'Toughening' of states' (literal and figurative) walls (Jones, 2012, 2021, 2022) has many consequences—many pungently deleterious—but it will not keep at bay the global harms that flow above, below, through and around borders. This is captured by Gilmore's (2023: 81) position that "the neat demarcation between inside and outside, if one ever really existed, is increasingly subverted by the transnationalisation of human relations and the pressing global problems that accompany them". By both (a) assigning human rights to *all* humans and (b) recognising the transboundary nature of harm, solidarism would deem communitarian particularism an impediment to the propagation of moral cosmopolitan values that seek to protect one human life as much as any other (Pogge, 1992: 48–49). Once more, please do not mistake me for a post-sovereign revolutionist: I reiterate that state-centric solidarism—even my more intrusive 'humankind' variant—is not a universalising project that attempts to usurp sovereignty and emancipate humans, but rather is one that takes existing power arrangements and sows the seeds of moral cosmopolitanism therein.

These thoughts have environmental implications. They lend themselves to Falkner's (2021: 67; emphasis added) understanding of global environmental governance, which entails constructing "strong *international* institutions based on a global consensus around *humanity's* universally shared environmental values and beliefs". The *institutions* are international; the *values* are universally shared by humanity. This speaks, in my mind at least, to Weinert's (2017, 2018, 2021, 2022) rich work on heritage, care, and humankind's shared geographies. Solidarism would propagate ambitious progress—in our case, concern for environmental accountability—within, as well as beyond, domestic jurisdictions. Universal(isable) concern for *all* humans' environments would be codified in an ambitious repertoire of law, whose legal "teeth" would "bite" inside borders. To the extent that a green solidarism would propagate universal(isable) values in a residually statist world, it would help reconcile Brundtland's (World Commission on Environment and Development, 1987: 27) dilemma that "the Earth is one but the world is not".

Such reconciliation of "one Earth" and the fractured political world could not come too soon, for solidarists. Hurrell (1994: 149) beckoned Falkner's call for "strong institutions" three decades beforehand. In a stroke of genius, which I regularly cite in the classroom, Hurrell (2007: 216) deemed the state concomitantly "too big" and "too small" to handle environmental challenges: "too big for the task of devising viable strategies of sustainable development which can only be developed from the bottom up; and too small for the effective management of global problems" such as climate change. The logic is that sovereignty needs reform, in order to be and to remain capable of tackling the most pressing environmental risks. This logic was evinced by Linklater's (2009: 489) extraction, from the ecological challenge, of opportunities "for reshaping moral and political horizons" in light of global harms that elude state control. The global environmental challenge is so complex, so subversive of borders, that pluralist minimalism is not enough. It needs to be ethically enriched with something *more*.

This demand for something *more*—more ambitious, more demanding of sovereignty, more willing to agitate states' ethical thresholds, is met by solidarisation (Ahrens, 2017, 2018, 2019; Ahrens & Diez, 2015), a fresh development that understands solidarism as *process*. Solidarisation offers a tool with which to bridge pluralism and solidarism. No longer need pluralism and solidarism to be understood as separate ideal types. The beauty of solidarisation lies in its capacity to portray pluralism and solidarism in a state of mutual interplay: both of these approaches are cast as mutually reinforcing and disciplining.

Solidarisation is a process wherein orthodox, statist frameworks are infused with more "humankind" ethical ambition. I refer to such infusion as the *cosmopolitan impulse*. It is intended to capture Aarhus' moral cosmopolitan ethical drivers: concerns for codifying human empowerment and "voice"; enhancements to state proclivities to "hear" such "voice"; state-human co-creation; and "humankind" governance modalities. This sense of infusion, of the cosmopolitan impulse ethically enriching sovereignty, denotes that pluralist, statist arrangements can be reformed *from within*. The qualities of (responsible democratic) states—humane treatment of individuals and associations "within the border"—would be propagated inside the existing international power arrangements of pluralism. A more humane, inclusive disposition towards non-nationals as well as co-nationals would be sought. This requires, for Ahrens (2019: 266), realignment of sovereignty "in terms of…more far-reaching definition of responsibilities and duties of states towards each other and vis-à-vis individuals inside and outside their own territories". Such realignment echoes Linklater's (2006: 124) idea of responsible statehood, defining sovereignty "in the light of cosmopolitan ideals". Rather than implying post-sovereign transformation, solidarisation is a cosmopolitanism of "little steps" (Archibugi, 2004: 466): a realistic cosmopolitanism (Beck & Sznaider, 2006: 8) wherein statism and cosmopolitanism help one another achieve their agendas. Auspiciously for the following chapters, Beck and Sznaider (2006: 9) regard realistic cosmopolitanism as emanating "from within" state structures. Such incrementalism constitutes "*internalised* cosmopolitanism…arising *inside* nation-states" (Beck & Sznaider, 2006: 9; emphasis added). The cosmopolitan impulse concomitantly constitutes (a) stimulus of ethical progress and agitator of states' ethical thresholds whilst (b) being confined and contained by existing power frameworks. Pluralism and solidarism materialise as symbiotic.

It is favourable, for the forthcoming chapters, that an indicator of solidarisation is the increasing "*procedural* involvement" of nonstate actors (Ahrens, 2017: 6; emphasis added). This may materialise in the casting of humans and associations as both "reference point…and as legitimate claimants of rights" (Ahrens, 2017: 7). Solidarisation implies the possibility that pluralist arrangements can be rendered more other-oriented and ethically ambitious; that incremental progress can occur in IR without "system change". This gels with Gilmore's (2023: 94) call to "forge a pluralist path to a more legitimate solidarist international society". Such employment of pluralism for solidarist ends speaks to the possibility of embedding ethical ambition *within* inherently pluralist frameworks, whilst retaining sovereignty as a disciplining influence on such ambition. States remain the gatekeepers of progress in IR. Interplay thus exists "between a transformative impetus and the necessity to engage with existing state-centric structures" (Ahrens, 2018: 3). The logic is that progress can only be achieved if sufficiently embedded in existing power arrangements, and if sufficiently palatable for consenting states. Such interplay between *reaffirmation* and *reform* of sovereignty is "not a detriment to change, but its explicit precondition" (Ahrens, 2018: 43).

Such pluralist–solidarist interplay has been assessed through environmental lenses. Falkner and Buzan (2019: 141; emphasis added) explored how the primary institution of environmentalism was admitted into international society: initially radical green rallying cries were "*channelled*" into state-centric political arrangements, in order to retain their palatability for states. Sovereignty was reinforced and legitimised; states were kept "on side". As Falkner and Buzan (2019: 142–143) remind us, green civil society campaigning constituted a "major demand factor behind the supply of state-centric international regulation". Radical impulses for change in fact reinforced states as guarantors of environmental protection and "*boosted* state-centric environmental solidarism" (Falkner & Buzan, 2019: 142–143; emphasis added).

This underscores the salience of state consent in international environmental politics and law. Consenting states 'give their blessing' to being ethically enriched and challenged, but only to the extent that such challenge remains acceptable for them. We will witness, in the subsequent chapters, states iterating their sovereignty and reaffirming their interests, from the Convention's propagation through to its eventual germination and growth. We will witness, through Aarhus' lifespan to date, the mutual regulation of Aarhus' solidarist ambitions and pluralist actualities. Pluralism and solidarism will indeed be seen to be "complementary, interconnected and interdependent" (Bain, 2021: 105). To cast them as symbiotic is to acknowledge that the pillars of pluralism—diversity, coexistence, order, sovereignty and the traditional power channels of law and diplomacy—can *incubate* and *safely contain* solidarist progress, whilst at one and the same time being reformed by it (Gilmore, 2023: 81).

Conclusion

The landscape that I just painted portrays three imaginaries of IR: a realist anarchy of state survival; a revolutionist cosmopolis of human emancipation; and a rationalist international society wherein states deliberately converge to lesser or greater degrees. The pluralist variant was cast as more communitarian in upholding the diversity of sovereign prerogatives. The solidarist variant, employed in this book, was cast as being motivated by a moral cosmopolitanism centring on the inherent value of human life, and the feasibility of fostering universal(isable) values in the residual context of sovereignty.

The three following chapters—on Aarhus' propagation (Chapter 3), germination (Chapter 4) and growth (Chapter 5)—elicit only slim evidence of realism and revolutionism, and are framed comfortably in the rationalist understanding of states' convergence, in our case upon the shared issue of environmental accountability. The following chapters will derive value from the concept of solidarisation. What follows is an observation of a potentially radical international environmental agreement being driven by an impulse of "humankindness": of an ostensibly mainline UN agreement being driven by well-codified concerns for human empowerment, voice and co-creation. These endeavours will be seen to ethically agitate sovereignty, whilst also being counterbalanced by it, in order to prevent "runaway" post-sovereign transformation. The following study of the Aarhus Convention will shed further light on the above solidarisation logic, and will lend further credence to my position that post-sovereign prospects are slim, given states' ability—and need—to contain potentially radical ethical progress in the pursuit of international order.

Bibliography

Agnew, J. (1994). The territorial trap: The geographical assumptions of IR theory. *Review of International Political Economy, 1*(1), 53–80.

Agnew, J., & Corbridge, S. (1995). *Mastering space: Hegemony, territory and international political economy*. Routledge.

Ahrens, B. (2017). *The solidarisation of international society: The EU in the global climate change regime* (GLOBUS Research Paper 5/2017). University of Oslo.

Ahrens, B. (2018) *AmbigEUity: the EU and the solidarisation of international society* (PhD thesis), Eberhard Karls Universität Tübingen.

Ahrens, B. (2019). The European Union between solidarist change and pluralist re-enactment In T. Brems Knudsen & C. Navari (Eds.), *International organisation in the anarchical society: The institutional structure of world order* (pp. 265–292). Palgrave Macmillan.

Ahrens, B., & Diez, T. (2015). Solidarisation and its limits: The EU and the transformation of international society. *Global Discourse, 5*(3), 341–355.

Ancel, J. (1936). *Géopolitique*. Delagrave.

Archibugi, D. (2004). Cosmopolitan democracy and its critics: A review. *European Journal of International Relations, 10*(3), 437–473.

Archibugi, D., & Held, D. (Eds.). (1995). *Cosmopolitan democracy.* Polity.

Armstrong, D., Lloyd, L., & Redmond, J. (2004) *International organisation in world politics.* Palgrave.

Bain, W. (2021). Pluralism and solidarism. In C. Navari (Ed.), *International society* (pp. 95–108). Palgrave.

Ball, T. (2006). Democracy. In A. Dobson & R. Eckersley (Eds.), *Political theory and the ecological challenge* (pp. 131–147). Cambridge University Press.

Barnett, J. (2003). Security and climate change. *Global Environmental Change, 13*(1), 7–17.

Beck, U. (1999). *World risk society.* Polity.

Beck, U., & Sznaider, N. (2006). Unpacking cosmopolitanism for the social sciences: A research agenda. *British Journal of Sociology, 57*(1), 1–23.

Berglund, O., & Schmidt, D. (2020). *Extinction rebellion and climate change activism: Breaking the law to change the world.* Palgrave Macmillan.

Booth, K. (1991). Security and emancipation. *Review of International Studies, 17*(4), 313–326.

Booth, K. (1997). Security and self: Reflections of a fallen realist. In K. Krause & M. C. Williams (Eds.), *Critical security studies: Concepts and cases* (pp. 83–119). UCL Press.

Booth, K. (2007). *Theory of world security.* Cambridge University Press.

Brems Knudsen, T. (2019). Fundamental institutions and international organisations: Solidarist architecture. In T. Brems Knudsen & C. Navari (eds.), *International organisation in the anarchical society: The institutional structure of world order* (pp. 175–202). Palgrave Macmillan.

Brown, C. (1992). *International relations theory: New normative approaches.* Columbia University Press.

Bull, H. (1981). Hobbes and the international anarchy. *Social Research, 48*(4), 717–738.

Bull, H. (2002). *The anarchical society: A study of order in world politics* (3rd ed.). Palgrave.

Buzan, B. (2004). *From international to world society? English school theory and the social structure of globalisation.* Cambridge University Press.

Buzan, B. (2014). *Introduction to the English school of international relations: The societal approach.* Polity.

Buzan, B., Waever, O., & de Wilde, J. (1998). *Security: A new framework for analysis.* Lynne Rienner.

Castells, M. (1996). *The information age: The rise of the network society* (Vol. 1). Blackwell.

Castells, M. (1997). *The information age: The power of identity* (Vol. 2). Blackwell.

Clark, L. C. (2022). Delivering life, delivering death: Reaper drones, hysteria and maternity. *Security Dialogue, 53*(1), 75–92.

Claude, I. (1984). *Swords into plowshares: The problems and progress of international organisation*. McGraw Hill.

Connelly, J. (2006). The virtues of environmental citizenship. In A. Dobson & D. Bell (Eds.), *Environmental citizenship* (pp. 49–74). MIT Press.

Convention on Access to Information, Public Participation in Decision-making and Access to Justice in Environmental Matters. (1998). Available at: https://unece.org/DAM/env/pp/documents/cep43e.pdf. Accessed 9 March 2023.

Curzon, G. (1907). *Frontiers*. Clarendon.

Dalby, S. (2000). Jousting with Malthus' ghost: Environment and conflict after the Cold War. *Geopolitics, 5*(1), 165–175.

Dobson, A. (2003). *Citizenship and the environment*. Oxford University Press.

Dobson, A. (2006). Citizenship. In A. Dobson & R. Eckersley (Eds.), *Political theory and the ecological challenge* (pp. 216–231). Cambridge University Press.

Donnelly, J. (1998). Human rights: A new standard of civilisation? *International Affairs, 74*(1), 1–24.

Dunne, T. (1998). *Inventing international society: A history of the English School*. Palgrave Macmillan.

Eckersley, R. (2006). Communitarianism. In A. Dobson & R. Eckersley (Eds.), *Political theory and the ecological challenge* (pp. 91–108). Cambridge University Press.

Epp, R. (2017). The British Committee on the theory of international politics and central figures in the English school. In N. Sandal (Ed.), *Oxford research encyclopaedia of international studies*. Available at: https://doi.org/10.1093/acrefore/9780190846626.001.0001/acrefore-9780190846626-e-562;jsessionid=6408175D8D5AFA7A0B93BB5D8863CB66?rskey=95G1CZ&result=145. Accessed 10 March 2023.

Falkner, R. (2012). Global environmentalism and the greening of international society. *International Affairs, 88*(3), 503–522.

Falkner, R. (2021). *Environmentalism and global international society*. Cambridge University Press.

Falkner, R., & Buzan, B. (2019). The emergence of environmental stewardship as a primary institution of global international society. *European Journal of International Relations, 25*(1), 131–155.

Flockhart, T., & Paikin, Z. (Eds.). (2023). *Rebooting global international society: Change, contestation and resilience*. Palgrave.

Gaddis, J. (1987). *The long peace: Inquiries into the history of the Cold War*. Oxford University Press.

Giddens, A. (1987). *Nation-state and violence*. Polity.

Gilmore, J. (2023). Conceptualising good global statehood: Progressive foreign policy after the populist movement. *International Theory, 15*(1), 79–105.

Gleick, P. H. (1991, April). Environment and security: The clear connections. *Bulletin of the Atomic Scientists*, 17–21.
Goldman Environmental Foundation. (2023). *1997 Goldman Prize Winner: Alexander Nikitin*. Available at: https://www.goldmanprize.org/recipient/alexander-nikitin/. Accessed 12 April 2023.
Hague, R., & Harrop, M. (2007). *Political science: A comparative introduction*. Palgrave Macmillan.
Harman-Stokes, K. M. (1995). Community right-to-know in the newly independent states of the former Soviet Union: Ending the culture of secrecy surrounding the environmental crisis. *Virginia Environmental Law Journal, 15*, 77–138.
Hartshorne, R. (1939). *The nature of geography: A critical survey of current thought in the light of the past*. Association of American Geographers.
Haushofer, K. (1932). *Wehr-Geopolitik: geographische Grundlagen einer Wehrkunde*. Junker und Dünnhaupt.
Held, D. (Ed.). (1993). *Prospects for democracy*. Polity
Holdich, T. H. (1916). *Political frontiers and boundary making*. Macmillan.
Homer-Dixon, T. (1994). Environmental scarcities and violent conflict: Evidence from cases. *International Security, 19*(1), 5–40.
Homer-Dixon, T. (1999). *Environment, scarcity and violence*. Princeton University Press.
Honneland, G. (2003). *Russia and the West: Environmental cooperation and conflict*. Routledge.
Huckle, J., & Sterling, S. (1996). *Education for sustainability*. Earthscan.
Hurrell, A. (1994). A crisis of ecological viability? Global environmental change and the nation state. *Political Studies, XLII*, 146–165.
Hurrell, A. (2007). *On global order*. Oxford University Press.
Hurrell, A., & Kingsbury, B. (Eds.). (1992). *The international politics of the environment: Actors, interests and institutions*. Clarendon Press.
International Commission on Intervention and State Sovereignty. (2001). *The responsibility to protect: Report of the International Commission on Intervention and State Sovereignty*. International Development Research Centre.
Jackson, R. (2003). *The global covenant: Human conduct in a world of states*. Oxford University Press.
James, A. (1986). *Sovereign statehood: The basis of international society*. Allen & Unwin.
Jones, R. (2012). *Border walls: Security and the war on terror in the United States, India, and Israel*. Zed Books.
Jones, R. (2021) *White borders: The history of race and immigration in the United States from Chinese exclusion to the border wall*. Beacon Press.
Jones, R. (2022). *Nobody is protected: How the border patrol became the most dangerous police force in the United States*. Counterpoint Press.

Kaplan, R. (1994,February). The coming anarchy. *Atlantic Monthly*, 44–76.

Kjellen, R. (1915). *Die Großmächte der Gegenwart*. Teubner.

Koliopoulos, C. (2019). International relations and the study of history. In N. Sandal (Ed.), *Oxford research encyclopaedia of international studies*.

Krause, K., & Williams, M. (Eds.). (1997). *Critical security studies: Concepts and cases*. UCL Press.

Laferrière, E., & Stoett, P. J. (1999). *International relations and ecological thought: towards a synthesis*. Routledge.

Lijphart, A. (1999). *Patterns of democracy*. Yale University Press.

Linklater, A. (1998). *The transformation of political community*. Polity.

Linklater, A. (2002). Cosmopolitan political communities in international relations. *International Relations, 16*(1), 135–150.

Linklater, A. (2006). Cosmopolitanism. In A. Dobson & R. Eckersley (Eds.), *Political theory and the ecological challenge* (pp. 109–127). Cambridge University Press.

Linklater, A. (2007). Towards a sociology of global morals with an emancipatory intent. *Review of International Studies, 33*, 135–150.

Linklater, A. (2009). Human interconnectedness. *International Relations, 23*(3), 481–497.

Linklater, A., & Suganami, H. (2006). *The English school of international relations: A contemporary reassessment*. Cambridge University Press.

Mackinder, H. (1904). The geographical pivot of history. *Geographical Journal, 23*(4), 421–437.

Marx, K., & Engels, F. (2002). *The Communist Manifesto*. Penguin.

Mearsheimer, J. (1990). Back to the future: Instability in Europe after the Cold War. *International Security, 15*(1), 5–56.

Morgenthau, H. (1945). The evil of politics and the ethics of evil. *Ethics*, 56(1), 1–18.

Morgenthau, H. (1985). *Politics among nations: The struggle for power and peace*. Knopf.

Myers, N. (1993). *Ultimate security*. Norton.

Navari, C. (Ed.). (2021a). *International society: The English school*. Palgrave.

Navari, C. (2021b). *The international society tradition: From Hugo Grotius to Hedley Bull*. Palgrave.

Paasi, A. (1999). Boundaries as social processes: Territoriality in the world of flows. In D. Newman (Ed.), *Boundaries, territory and postmodernity*. Cass.

Paasi, A. (2001). Europe as a social process and discourse: Considerations of place, boundaries and identity. *European Urban and Regional Studies, 8*(1), 7–28.

Paasi, A. (2009). Bounded spaces in a "borderless world": Border studies, power and the anatomy of territory. *Journal of Power, 2*(2), 213–234.

Pogge, T. (1992). Cosmopolitanism and sovereignty. *Ethics, 103*(1), 48–75.

Press Service of the President of the Republic of Belarus. (2022). *Belarus to withdraw from Aarhus Convention*. Available at: https://president.gov.by/en/events/belarus-to-withdraw-from-aarhus-convention. Accessed 6 May 2023.
Ratzel, F. (1897). *Politische Geographie*. Oldenbourg.
Rengger, N. J. (1996). A city which sustains all things? Communitarianism and international society. In R. Fawn & J. Larkins (Eds.), *International society after the Cold War: Anarchy and order reconsidered* (pp. 63–83). Macmillan.
Roberts, A., & Kingsbury, B. (Eds.). (1993). *United Nations, divided world: The UN's roles in international relations*. Clarendon.
Rosecrance, R. N. (1966). Bipolarity, multipolarity and the future. *Journal of Conflict Resolution, 10*(3), 314–327.
Shapcott, R. (2008). Anti-cosmopolitanism, pluralism and the cosmopolitan harm principle. *Review of International Studies, 34*, 185–205.
Schmitt, C. (2003). *The nomos of the Earth in the international law of the Jus Publicum Europaeum* (G. L. Ulmen, Trans.). Telos.
Schmitt, C. (2005). Political theology: Four chapters on the concept of sovereignty (G. Schwab, Trans.). University of Chicago Press.
Schmitt, C. (2007). The concept of the political (G. Schwab, Trans.). University of Chicago Press.
Selcer, P. (2018). *The postwar origins of the global environment: How the United Nations built Spaceship Earth*. Columbia University Press.
Suganami, H. (1983). The structure of institutionalism: An anatomy of British mainstream international relations. *International Relations, 7*(5), 2363–2381.
Taylor, P. (1994). The state as container: Territoriality in the modern world-system. *Progress in Human Geography, 18*(2), 151–162.
Universal Declaration of Human Rights. (1948). United Nations General Assembly, A/811, 16 December 1948.
Vincent, R. J. (1981). The Hobbesian tradition in twentieth century international thought. *Millennium, 10*(2), 91–101.
Wall, D. (2002). *Earth First! And the anti-roads movement*. Routledge.
Waltz, K. (1979). *Theory of international politics*. McGraw-Hill.
Warf, B. (2009). From surfaces to networks. In B. Warf & S. Arias (Eds.), *The spatial turn: Interdisciplinary perspectives* (pp. 59–76). Routledge.
Walzer, M. (1994). *Thick and thin: Moral argument at home and abroad*. University of Notre Dame Press.
Weaver, D. (2020). Spatiality and world politics. In N. Sandal (Ed.), *Oxford research encyclopaedia of international studies*.
Weaver, D. (2022). Extinction Rebellion: Greening vanguardism? *Social Movement Studies*.
Weinert, M. (2011). Reframing the Pluralist-Solidarist Debate. *Millennium, 40*(1), 21–41.

Weinert, M. (2017). Grounding world society: Spatiality, cultural heritage, and our world as shared geographies. *Review of International Studies, 43*(3), 409–429.

Weinert, M. (2018). Reading world society phenomenologically: An illustration drawing upon the cultural heritage of humankind. *International Politics, 55*, 26–40.

Weinert, M. (2021). Cultural heritage, cultural rights and care ethics. In H. Mahmoudi, A. Brysk, & K. Seaman (Eds.), *The changing ethos of human rights* (pp. 136–156). Edward Elgar.

Weinert, M. (2022). Integrating cultural heritage into human security analysis. *Global Governance, 28*(1), 35–57.

Wheeler, N. (2000). *Saving strangers: Humanitarian intervention in international society*. Oxford University Press.

Wight, M. (1991). *International theory: The three traditions*. Leicester University Press.

Williams, J. (2013). The international society—World society distinction. In C. Navari & D. Green (Eds.), *Guide to the English school in international studies* (pp. 127–142). Wiley.

World Commission on Environment and Development. (1987). *Our common future*. Oxford University Press.

Ziegler, C. E. (1987). *Environmental policy in the USSR*. University of Massachusetts Press.

CHAPTER 3

Propagation: The Aarhus Convention's International Context

Introduction

Readers might be forgiven for deeming Aarhus a revolutionary pioneer in international environmental politics and law, springing upon the world a radical new suite of rights. After all, it played lead character in Barritt's (2020: 12–17) legal "love story". But we must note Aarhus' situation in a large corpus of legal, diplomatic and geopolitical context. It did not emerge without warning. It was the outcome of decades' statecraft, setting the conditions in which its solidarist seeds could be propagated. It is possible to trace a trajectory, from the end of the Second World War onwards, towards the internationalisation of solidarist concerns for deeper, broader state convergence on human rights and environmental protection. But we must not naïvely deem this postwar era "peaceful". Whilst the UN was born, with human rights and environmental protection receiving their rightful place on the international agenda, the postwar era was marred by conflict (see Calvocoressi, 2013 for what remains, in my mind, the most engaging and comprehensive account; see also Lundestad, 2017). Perilous nuclear and ideological tensions existed in the latter half of the twentieth century, a time known as the Cold War given the formal absence of "hot" direct military engagement between two superpower blocs that coalesced around (a) the US and NATO and (b) the USSR and Warsaw Pact.

D. Weaver, *The Aarhus Convention*, Environmental Politics and Theory, https://doi.org/10.1007/978-3-031-43536-2_3

What was evident, in the postwar era, was a quest for an international order that would avoid a resurgence of global war, and which would secure the conditions for nonviolent coexistence. This was states' key focus after 1945: on preserving sovereignty, and the corollary principles of non-intervention and territorial integrity, thereby enacting the principle of "never again". Pursuit of more solidarist, other-oriented concerns was auxiliary to the pluralist quest for international order. There is a sense, in what follows, that the solidarist values of human rights and environmental protection could only be achieved if stable international order was achieved. Absent *order*, and notions of human and environmental *justice* would remain a mirage, ever eluding those who sought it. Such a pluralist concern for order does not necessarily imply the risk of a breakdown into war, but rather the risk that "radically unbalanced power will permit the powerful to "lay down the law" to the less powerful, to skew the terms of cooperation in its own favour, to impose its own values and ways of doing things, and to undermine the procedural rules on which stable and legitimate cooperation must inevitably depend" (Hurrell, 2007: 289). Such destabilisation would no doubt aggravate the risks of global war in a world already marred by two such conflicts.

With this in mind, there is a sense, in the following analysis, that solidarist ideals would (a) reform and challenge states' practices whilst (b) being confined by sovereignty and the quest for coexistence. To draw these findings, the chapter takes the following steps. It firstly assesses the legal and diplomatic context enabling Aarhus' propagation, specifically concerning human rights and environmental protection. Secondly, it addresses the geopolitical context, making the case that Aarhus would not have emerged without the democratisation, "East–West" dialogue, and rise of socialist civil society that preceded and accompanied the Cold War's end. By addressing those contexts the chapter can, thirdly, assess the multilateral Environment for Europe initiative, under whose auspices Aarhus was propagated. Environment for Europe is notable, as a multilateral process, for the gratifying degree to which it sought an amplification of human environmental voice, and ensured that such voice was "heard" by states. The chapter finishes by assessing the role of democracies and post-socialist states during Aarhus' launch, before distilling its findings.

International Legal and Diplomatic Context

Human Rights

United Nations Charter and 'International Bill of Rights'

Aarhus was negotiated, signed and is implemented, under United Nations Economic Commission for Europe's (UNECE's) auspices. It is thus prudent to contextualise its propagation with reference to the UN. We begin with the UN's founding document, the UN Charter (Charter of the United Nations & Statute of the International Court of Justice, 1945). This sought to preserve international order after two world wars. It attempted to stabilise IR, given the League of Nations' failure (Lloyd, 1995). The Charter preserves the sanctity of sovereignty and seeks to avoid unnecessary incursions into states' domestic jurisdictions. This is captured in the provision that "Nothing contained in the present Charter shall authorize the United Nations to intervene in matters which are essentially within the domestic jurisdiction of any state" (Charter of the United Nations & Statute of the International Court of Justice, 1945, Article 2.7). It is also captured in the requirement that states seek nonviolent, conciliatory dispute settlement (Charter of the United Nations & Statute of the International Court of Justice, 1945, Article 33), with force only to be authorised by the UN Security Council if "measures not involving the use of armed force" (Charter of the United Nations & Statute of the International Court of Justice, 1945, Article 41) "would be inadequate or have proved to be inadequate" (Charter of the United Nations & Statute of the International Court of Justice, 1945, Article 42). Safe coexistence is sought, embodying a "return of the old notion of jus ad bellum and the view that the use of force was only to be justified in case of self-defence or as authorised by the UN" (Hurrell, 2007: 182; Charter of the United Nations & Statute of the International Court of Justice, 1945, Articles 39–50 on Security Council authorisation of force; Article 51 on self-defence). It should be agreed that the UN "was created by states for states" (Curtis & Taylor, 2017: 332). Sovereign equality (Charter of the United Nations & Statute of the International Court of Justice, 1945, Article 2.1) is key: each member state, whatever its politics, has a formally equal place in "the essential club for states" (Curtis & Taylor, 2017: 344). This resonates with pluralism, as it matters not whether a state is large or small, strong or weak, rich or poor: what matters is that they have—in theory, at least—no greater or lesser flag on the "world stage". Diversity and difference are preserved; a "live and let live" ethic is espoused.

Yet there are glimpses of solidarism in the Charter. Its Preamble (Charter of the United Nations & Statute of the International Court of Justice, 1945) reaffirms "faith in fundamental human rights, in the dignity and worth of the human person, in the equal rights of men and women". The Preamble (Charter of the United Nations & Statute of the International Court of Justice, 1945, emphasis added) seeks "to establish conditions under which *justice* and respect for the obligations arising from treaties and other sources of international law can be maintained, and to promote *social progress* and better standards of life in *larger freedom*". The UN's purpose was to "achieve international cooperation in solving international problems of an economic, social, cultural, or humanitarian character, and in promoting and encouraging respect for human rights and ...fundamental freedoms for all" (Charter of the United Nations & Statute of the International Court of Justice, 1945, Article 1.3). As such, the Charter navigates between (a) sanctifying, in pluralist tones, sovereignty and non-intervention whilst (b) aspiring to the solidarist ideals of deeper convergence, human rights and "fundamental freedom". It is as though *justice* cannot be contemplated in the absence of *order*. The pluralist prerogatives materialise as prerequisites for solidarist ambition. The ethically exacting pursuit of human (and environmental) justice will simply not be feasible without an underlying bedrock of nonviolent, orderly international coexistence. Remove that bedrock, remove that nonviolent stability, and the ensuing chaos would reduce to fantasy any chances of achieving the goal of human (and environmental) justice.

Three and a half years after the Charter's publication, the Universal Declaration of Human Rights (1948) was issued by the UN General Assembly. It was the first of a trilogy of efforts constituting an international "bill of rights". Whilst the Universal Declaration is aspirational, its two counterpart documents are binding (International Covenant on Civil and Political Rights, 1966; International Covenant on Economic, Social and Cultural Rights, 1966; see HM Parliament, 2004: 5). The Universal Declaration evinces solidarist ambition, confined in the architectures of pluralist order. This is evident in the provision (Universal Declaration of Human Rights 1948, Article 28) that every human is "entitled to a social and international order in which the rights and freedoms set forth in this

Declaration can be fully realised". Article 28 reflected the drafting authorities' view that "a world at peace was essential for respect for human rights and to create opportunities for everyone to improve their lives" (Office of the High Commissioner for Human Rights, 2018, online). Unless order was secured, solidarist ideals would be unattainable.

The Universal Declaration's (Universal Declaration of Human Rights, 1948, Article 1; emphasis added) first operative provision does not shy away from solidarist aspiration. It states that "All human beings are born free and equal in dignity and rights. They are endowed with reason and conscience and should act towards one another in a *spirit of brotherhood*". Procedural rights are addressed too. The Universal Declaration (Universal Declaration of Human Rights, 1948) provides for freedoms of thought, conscience and religion (Article 18), and of opinion and expression (Article 19). Noteworthy for our study is the latter's "freedom to hold opinions without interference and to seek, receive and impart information and ideas through any media and regardless of frontiers (Universal Declaration of Human Rights, 1948, Article 19). Public participation surfaces in the Declaration's rights to assembly and association (Universal Declaration of Human Rights, 1948, Article 20.1) and to "take part" in government "*directly* or through *freely chosen representatives*" (Universal Declaration of Human Rights, 1948, Article 21.1; emphasis added). This is complemented by the right "freely to participate in the cultural life of the community, to enjoy the arts and to share in scientific advancement and its benefits" (Universal Declaration of Human Rights, 1948, Article 27). Modest allusion to substantive environmental quality emerges in the Declaration's right to a "standard of living adequate for the health and well-being" of oneself and one's family (Universal Declaration of Human Rights, 1948, Article 25.1). But nonhuman nature is elusive. The Declaration is aligned closely with the UN Charter: it recognises that human justice can only be achieved in a world free of war, where states can begin to tentatively contemplate their international interests without recourse to violence.

The Declaration was reinforced by the binding (HM Parliament, 2004) International Covenant on Civil and Political Rights (1966) and International Covenant on Economic, Social and Cultural Rights (1966). A "close relationship" has been said to exist between Aarhus' provisions and those provided by the International Covenant on Civil and Political Rights (UNECE, 2005: 13; see UNECE, 2022: 17). The International Covenant on Civil and Political Rights (1966) reinforced not only the right to life (Article 6.1) but the human status of legal person (Article 16), the freedoms of thought, conscience and religion (Article 18), expression (Article 19), assembly (Article 21) and association (Article 22). Auspiciously for our study, it affirmed the right to participate in public affairs (Article 25a).

The International Covenant on Economic, Social and Cultural Rights (1966, Article 11.1), meanwhile, reinforced the Universal Declaration's allusion to living conditions, proclaiming the right "to an adequate standard of living…including adequate food, clothing and housing, and to the continuous improvement of living conditions". It recognised the right to "the highest attainable standard of physical and mental health" (Article 12.1) necessitating, inter alia, "improvement of *all aspects* of *environmental and industrial hygiene*" (Article 12.2.b; emphasis added). It reaffirmed the Universal Declaration's right to participation in scientific and cultural life (Article 15). As such, the Covenants helped to actualise the Universal Declaration's aspirations, propagating seeds of international order and human dignity that helped to enrich the earth on which Aarhus would later itself be sown.

The International Bill of Rights was supplemented and indeed preceded by ancillary efforts in "softer" law and diplomacy. Freedom of information, for instance, animated the first sitting of the UN General Assembly, some two years before the Universal Declaration was published. Resolution 59(I) (United Nations General Assembly, 1946) deemed freedom of information "a fundamental human right" and "the touchstone of all the freedoms to which the United Nations is consecrated". Resolution 59(I) regarded states' cooperation as "impossible without an alert and sound world opinion which, in turn, is wholly dependent upon freedom of information" (United Nations General Assembly, 1946). The Resolution proposed a conference on freedom of information, which occurred in 1948, but yielded "a victory of limited objectives" (Whitton, 1949: 73), not least given the emergence of East–West tensions with the onset of the Cold War. Three draft Conventions were adopted at the freedom of information conference: a "US Convention" on gathering and transmitting news; a "French Convention" on an international right of correction; and a "British Convention" on freedom of information principles (Whitton, 1949: 74). Only the "French" variant (Convention on the International Right of Correction, 1953) entered into force. Proposals for an international convention on freedom of information bounced around the UN for a very long time, and to little avail (as a snapshot from the labyrinthine archives, see United Nations General Assembly, 1962, 1974). Indeed, with the exception of Aarhus, I would argue that freedom of information remains relatively underrepresented in universally codified international law. Whilst the right is protected in the European human rights system, and whilst it is the subject of significant European Court of Human Rights (ECtHR) case law, at a global scale it continues to be derived from the Universal Declaration (United Nations Educational, Scientific and Cultural Organisation, 2022). The Tromsø Convention (Council of Europe Convention on Access to Official Documents, 2008), potentially a very valuable innovation in this domain, has but 15 ratifications at the time of writing (Council of Europe, 2023), and is thus by no means a runaway success.

Similarly, the General Assembly's (1969) Declaration on Social Progress and Development constituted a nonbinding supplement to the Bill of Rights. It is a shame that this aspiration appears to have collected dust in the archives (but see Kiwanuka, 1988), for it demonstrates the tensions, at that time, between the priorities of development, environmental protection and procedural empowerment. Article 5 (United Nations General Assembly, 1969) speaks to the procedural rights of information access and public participation, encouraging "creative initiative under conditions of enlightened public opinion", dissemination of information "for the purpose of making individuals aware of changes occurring in society" and the "active participation of all elements of society" (see also Article 15). But whilst this aspiration is clearly not environment-blind (Articles 13c and 25a), it betrays an industrial hunger that would unnerve environmentalists, and which was not aligned with later concerns for sustainable development (see below). Article 17 (United Nations General Assembly, 1969) seeks an arguably incommensurable balance between "measures to accelerate the process of industrialisation, especially in developing countries" and "measures designed to ensure healthier living conditions, especially in large industrial centres". The balancing act also surfaces in Article 23's (United Nations General Assembly, 1969; emphasis added) provision "to facilitate the *direct exploitation* of [developing countries'] national resources and natural wealth". The rhetoric is blurred, colonising and extractive, juxtaposing as it does industrial hunger and "direct exploitation" against a "bolt on" secondary concern for living conditions. Such tensions in development continued, and were embodied not least by the malaise of the Millennium Development Goals

(MDGs) and subsequent struggle towards the Sustainable Development Goals (SDGs) (see Briant Carant, 2017).

The European Convention on Human Rights
Two years after the Universal Declaration, the European Convention on Human Rights (ECHR) (Convention for the Protection of Human Rights & Fundamental Freedoms, 1950) was opened, entering into force in 1953. The ECHR is a binding Council of Europe instrument; it is worth recalling that the Council of Europe is a distinct international organisation, separate from the European Union (EU), established in 1949 to create "a common democratic and legal area...ensuring respect for human rights, democracy and the rule of law" (Heywood, 2014: 490). The ECHR was the first binding instrument to codify some of the Universal Declaration's provisions. It is enforced by the ECtHR, "a long-standing institution, argued to be the world's most effective international tool for the protection of human rights" (Brown & Ainley, 2009: 234). In this context, the ECHR has been deemed "the nearest thing to human rights 'hard' law" (Heywood, 2014: 319).

The ECHR does not contain substantive environmental provisions (Darpö, 2018; Peters, 201813): environmental rights have to be derived from it (Birnie et al., 2009: 300). Some commentators thus regarded the ECtHR as having had to develop "a tradition of steering that instrument [ECHR] towards more substantial environmental protections" (Braig et al., 2022: 74). For this reason, the ECHR is problematic in that environmental litigation will only "reach" the ECtHR if an ECHR substantive right is infringed. In other words, the ECtHR needs a "causal link between an instance of environmental degradation and an infringement of the Convention" in circumstances "where there is not only a mere risk but also a concrete threat to individual rights" codified by the ECHR (Braig et al., 2022: 80, emphasis added; see Eicke, 2022; Peters, 2018). The result is a form of environmental litigation, in ECtHR, that tends to focus on the ECHR's (Convention for the Protection of Human Rights & Fundamental Freedoms, 1950) substantive provisions for private and family life (Article 8), life itself (Article 2) and property (Protocol to the Convention for the Protection of Human Rights & Fundamental Freedoms, 1952, Article 1). This is noteworthy, as there is no such precondition for substantive rights violations in Aarhus. Furthermore, scholars observed not only that the ECtHR cited Aarhus when interpreting the ECHR (Braig et al., 2022: 83), but that the European Court of Justice (ECJ) cited Aarhus in its own proceedings as well (Darpö, 2018). Indeed, Peters (2018: 22) suggested that the ECHR "allows far less individual involvement in environmental governance" than Aarhus.

A case should be made that the ECHR's *raison d'etre* was to stabilise order in Europe, whilst at the same time setting the framework in which humans could be treated fairly and in a state of postwar human rights compliance. The ECHR's initial purpose was less discretely environmental. More broadly, the above human rights instruments served largely as aspirational conduits through which *states* could navigate their competing priorities of stabilising international order, accelerating environmentally intrusive development, and setting a political trajectory that would preserve human dignity: something distinctly lacking by the time the Second World War had concluded.

Environmental Protection

Towards Stockholm: United Nations Conference on the Human Environment

Having explored the postwar human rights context, I will now assess the environmental milestones. We will shortly visit the United Nations Conference on the Human Environment (UNCHE) (United Nations, 1972) held in Stockholm. As we shall see, it was a gamechanger, a "watershed" (Elliott, 2004: 7) in the history of environmental multilateralism. But before arriving, let me stress that Stockholm built on previous efforts to place the environment on the agenda. As early as 1948, the UN arranged for the Scientific Conference on the Conservation and Utilisation of Resources (Gibboney, 1949; United Nations Economic & Social Council, 1948). This occurred alongside the International Technical Conference on the Protection of Nature (United Nations Educational, Scientific and Cultural Organisation, 1950). These early attempts were not alone. They were accompanied by efforts, for instance, to regulate whaling, high seas fisheries and Antarctic resource exploration (Falkner, 2021: 108–109). But, as Falkner (2021: 108) observes, these early efforts did not indicate "desire among the most powerful states to move beyond the limitations of the Green Westphalian approach". And those early efforts were, to join Elliott (2004: 8), "dominated by states and governments".

Modest progress came four years before Stockholm, with the Intergovernmental Conference of Experts on the Scientific Basis for Rational

Use and Conservation of the Resources of the Biosphere (the Biosphere Conference) (United Nations Economic & Social Council, 1969). It was held in Paris in 1968 under the auspices of United Nations Educational, Scientific and Cultural Organisation (UNESCO) (United Nations Educational, Scientific and Cultural Organisation, 2023 for valuable archival recordings). The Biosphere Conference occurred in the context of growing public concern for environmental protection, not least concerning the dangers of pesticides (Carson, 1962) and unregulated use of the commons (Hardin, 1968). The Biosphere Conference addressed the "changing biosphere and man's role as beneficiary, violator and manipulator of his environment" (United Nations Economic & Social Council, 1969: 10). NGOs worked with states and intergovernmental organisations (IGOs), setting participatory custom (United Nations Economic & Social Council, 1969: 35). The outcome was a research programme on the connections between humankind and the environment (United Nations Economic & Social Council, 1969: 36, 194), which still exists today in the form of the "Man and the Biosphere" programme (United Nations Educational, Scientific and Cultural Organisation, 2021).

Yet I claimed modest progress, at the top of the preceding paragraph. There is a sense in the archives that the Biosphere Conference yielded limited impact. The UN Economic and Social Council (1969: 195) recorded, a year later, persistent "cases of pollution of the sea by oil, lost atomic bombs and the holding of nuclear tests". In light of the ongoing environmental harm, the Economic and Social Council (1969: 195–198) invested hope in a proposed UN Conference on the Human Environment, with an official record noting that "all interested States should be invited to participate in it" and that "meaningful action" should result from it, as "the conference on the human environment would provide proof to the public of the effectiveness of United Nations action". Stockholm was to be a test of states' collective mobilisation on environmental protection.

Stockholm (United Nations, 1972) indeed consolidated the environment as a priority on the international agenda. It marked the "constitutional moment in the creation of environmental stewardship as a primary institution of international society" (Falkner, 2021: 118). Around 1200 representatives of over 100 states participated (Elliott, 2004: 11; United Nations, 1972: 43), plus those of the UN, IGOs and NGOs. Presciently, the General Assembly Resolution (1968) providing for UNCHE noted NGOs' role in environmental protection. This role was apparent at Stockholm. A formal NGO conference, plus an informal People's Forum, set valuable participatory precedent (Elliott, 2004: 11). As such, a cosmopolitan impulse of human empowerment, voice and state-human co-creation was propagated at Stockholm. This is substantiated in the opening statement (United Nations, 1972: 45; emphasis added) of UNCHE Secretary-General Maurice Strong, who "stressed the need for...*new concepts of sovereignty*, based not on the surrender of national sovereignties but on better means of *exercising them collectively*, and with a greater sense of *responsibility for the common good*". Furthermore, throughout Stockholm "it was repeatedly emphasised that it was only by means of global co-operation, understanding, sympathy and assistance that mankind could achieve the results that his talents could provide" (United Nations, 1972: 48). Green solidarist seeds were sown, in a way cognizant of the need for sovereignty's reform, half a century ago.

The resultant Stockholm Declaration sought to raise environmental consciousness amongst states without eroding the primacy of state sovereignty. It attempted to balance pluralist affirmations of sovereignty and greener, more ambitious other-orientation. Principle 24 (United Nations, 1972: 5; emphasis added) demonstrates this:

> International matters concerning the protection and improvement of the environment should be handled in a *co-operative spirit by all countries*, big and small, on an equal footing. Co-operation through multilateral or bilateral arrangements or other appropriate means is essential to effectively control, prevent, reduce and eliminate adverse environmental effects resulting from activities conducted in all spheres, in such a way that due account is taken of *the sovereignty and interests of all States*.

This is not to deny Stockholm's clear recognition that humankind was central to environmental protection. After all, humankind was cast by the Stockholm Declaration (United Nations, 1972: 3) as "both creature and moulder of his environment", with "the long and tortuous evolution of the human race" reaching a stage when "man has acquired the power to transform his environment in countless ways and on an unprecedented scale. Both aspects of man's environment, the natural and man-made, are essential to his well-being and to the enjoyment of basic human rights – even the right to life itself". As "creature", humankind enjoys myriad environmental benefits, unchallenged as its predatorial status is amongst the species. As "moulder", humankind harnesses, and yields dominion over, nonhuman nature. The corollary is that human and environmental health are symbiotic. If one or other is injured, humankind's capacity to protect, and benefit from, such rights—including the most fundamental, the right to life—is compromised. This is captured in the Stockholm Declaration's (United Nations, 1972: 3) statement that humankind's ability "to transform his surroundings, if used wisely, can bring to all peoples the benefits of development and the opportunity to enhance the quality of life. Wrongly or heedlessly applied, the same power can do incalculable harm". Stockholm thus drew two equations, between (a) human and environmental rights and (b) rights and responsibilities (United Nations, 1972: 4; Principle 1). Similar equations surface in Aarhus' Preamble (see this book, Chapter 1). Human agency materialises as a vehicle for improving environmental prospects (United Nations, 1972: 5, Principles 19 and 20), in recognition that "ignorance or indifference...can do massive and irreversible harm to the earthly environment" (United Nations, 1972: 3).

These aspirations hinged on state consent. Stockholm sought compromise between the competing interests of developed and developing states, the latter concerned that "first world problems" like conservation would be favoured over poverty (Elliott, 2004: 10–11). Stockholm's Action Plan for the Human Environment (United Nations, 1972: 6–28) was a "catalogue of concerns" (Elliott, 2004: 12). But it would be unfair to deem UNCHE a failure. It placed environmental protection irreversibly on the agenda, with states playing lead roles. UNCHE preceded growth in international environmental agreements (Falkner, 2021: 131); it stimulated the creation of no less than the United Nations Environment Programme (UNEP) itself (United Nations, 1972: 7). The Action Plan, however troublesome, was animated by procedural concern, not least for information access and knowledge exchange (United Nations, 1972: 8, 9, 24, 26). Environmental awareness was deemed a stimulus of "active participation by the citizens" (United Nations, 1972: 25). State capacities for human empowerment in environmental contexts were mobilised. Solidarist objectives were propagated "within an unambiguous reassertion of the principles of national sovereignty and development" (Falkner, 2021: 125).

Towards Rio: United Nations Conference on Environment and Development

The twenty years between Stockholm and the 1992 United Nations Conference on Environment and Development (UNCED, the Rio Earth Summit) (United Nations, 1993) were marked by a flurry of multilateralism (Falkner, 2021: 128–148). States sought to overcome "initial inertia" and build on Stockholm's momentum. Highlights, in this interregnum, included the first World Climate Conference in 1979 (Zillman, 2009); the Montreal Protocol on Substances that Deplete the Ozone Layer (1987), which is "to date one of the rare treaties to achieve universal ratification" (United Nations Environment Programme, 2023); and agreements on oceanic pollution and endangered species (Elliott, 2004: 13).

Other highlights set clearer procedural precedent for Aarhus. The Geneva Convention on Long-Range Transboundary Air Pollution (1979) set precedent for Aarhus (Fasoli & McGlone, 2018: 32) by spearheading procedural rights (Okowa, 1997: 300) to information exchange. The Espoo Convention on Environmental Impact Assessment in a Transboundary Context (1991) set precedent by enacting information and participation rights in circumstances of transboundary environmental harm (Convention on Environmental Impact Assessment in a Transboundary Context, 1991, Article 2.6, Article 3.8; see Ebbesson, 2011). Espoo complemented not only UNEP's (1987) Goals and Principles on Environmental Impact Assessment, but also European Community (EC) Directives 85/337 on environmental impact assessment, and 90/313 on environmental information access. Aarhus elaborated upon these latter EC directives (Blundell, 2015: 99; Cramer, 2009: 95; Hallo, 2011: 58; Jendrośka, 2005: 13; Pallemaerts, 2011: 6).

To repeat, I understand this post-Stockholm flurry as being driven by a collective international effort to maintain the momentum of Stockholm's progress. Concerns for procedural propriety were clearly evident. Take, for instance the General Assembly's (1982) adoption of the "World Charter for Nature", whose sixteenth provision requires "formulation of strategies for the conservation of nature, the establishment of inventories of ecosystems and assessments of the effects on nature of proposed policies and activities; all of these elements shall be disclosed to the public by appropriate means in time to permit effective consultation and participation". This participatory tone continued with the World Charter's twenty-first provision (United Nations General Assembly, 1982;

emphasis added) that "States and, to the extent they are able, other public authorities, international organisations, individuals, groups and corporations shall…co-operate in the task of conserving nature through *common activities* and other relevant actions, including *information exchange and consultations*". A push towards co-creation is evident: there is a recognition that states cannot "go it alone" in pursuing global environmental protection. There appears to be an attempted harnessing of human voice and reasoning for environmental purposes. The World Charter's penultimate provision (United Nations General Assembly, 1982; emphasis added) closes the document in a firm tone, stating that each person "shall have the opportunity to participate, individually or with others, in the formulation of *decisions of direct concern to their environment*, and shall have access to means of redress when their environment has suffered damage or degradation". This resonates with the all-affected principle, an ancient cosmopolitan principle requiring that "what touches all should be heard and approved by all" (see Linklater, 2002: 146). We will revisit this principle in the next chapter (Chapter 4).

Such a strengthening of the trajectory towards co-creation, procedural propriety and human empowerment indicates states' acceptance, in the mid-late twentieth century, that humans were deserving of knowledge, participatory opportunities and access to justice in circumstances where their, and their environments', health was at stake. It accompanied a broad global push, at that time, towards sustainable development: socioeconomic development that would not "cost the Earth". Nowhere is this clearer than *Our Common Future* (WCED, 1987), the eponymous Brundtland Report. WCED constituted a response to the UN's call for approaches to development that would be sustainable, both for humankind and the environment, in the long term. It set participatory precedent, given its use of worldwide public hearings allowing Commissioners to obtain diverse public views (Elliott, 2004: 14). I have no doubt that Brundtland was implicated in, and contributed to, the propagation of green solidarist seeds in international environmental politics and law, towards the end of the last century. Brundtland's famous definition of sustainable development, as that which "meets the needs of the present without compromising the ability of future generations to meet their own needs" (WCED, 1987: 43), more acutely inclines us towards the unknown, the unborn, the distant and those in "need". The term "Common", in its title, is significant: it is surely no coincidence that it follows the Brandt Commission on North–South Issues (incurring the Brandt

Programme for Survival and *Common* Crisis) and Palme Commission on Security and Disarmament (incurring *Common* Security) (WCED, 1987: 6). Commonality is the unifier: it places the shared environmental and socioeconomic vulnerabilities of all "united" nations as a global priority. Indeed, Brundtland (WCED, 1987: 6) identified, and cautioned against, the troubling risks of post-Stockholm recidivism:

> The present decade has been marked by a retreat from social concerns. Scientists bring to our attention urgent but complex problems bearing on our very survival... We respond by demanding more details, and by assigning the problems to institutions ill-equipped to cope with them...And as part of our 'development', we have amassed weapons arsenals capable of diverting the paths that evolution has followed for millions of years and of creating a planet our ancestors would not recognize.

This disquieting rhetoric, of particular pungency as I finish this book, is indicative of a concern, at that time, about environmental regression following UNCHE's initial progress. It is as though the environment had been instrumentalised and commodified, with 'humankindness' and environmental care being stifled by runaway "development", not only economic but also military. In my reading, Brundtland's logic was that development was inseparable from care. Sustainable development, cognizant of environmental and human vulnerability, warranted "changes in attitudes, in social values, and in aspirations that...depend on vast campaigns of education, debate and public participation" (WCED, 1987: 8–9). A strengthening of procedural rights was necessary, and the identification of this paved the way for the early calls for more formal international agreement on environmental transparency and accountability, a few years later. Environmental information was deemed a prerequisite for "an informed basis for public discussion" (WCED, 1987: 57). Environmental planning was said to require the "views of non-governmental organisations and the local community", whilst "necessary information should be disclosed to nearby residents in an easily understandable manner" (WCED, 1987: 192). Participatory precedent is found in Brundtland's (WCED, 1987: 270) finding that:

> governments need to recognize and extend NGOs' right to know and have access to information on the environment and natural resources; their right to be consulted and to participate in decision making on activities likely to have a significant effect on their environment; and their right to legal

remedies and redress when their health or environment has been or may be seriously affected.

This statement, five years before Rio Principle 10 (see below), is sufficient evidence with which to reinforce Falkner's (2021: 151) claim that Brundtland achieved "breakthrough" in advance of UNCED. A year afterwards, the Intergovernmental Panel on Climate Change was created; by 1989, environmental protection was on the G7 agenda (Elliott, 2004: 13–14). Brundtland was supplemented by such ambitions as the World Health Organisation's European Charter on Environment and Health (1989: 2), which declared rights to environmental information access and public participation. It was supplemented by a UN General Assembly (1990) Resolution on the need to ensure a healthy environment for the well-being of individuals. And it was augmented by a subsidiary of the Commission on Human Rights, the Sub-Commission on Prevention of Discrimination and Protection of Minorities, mandating a study of the links between human rights and environmental harm (United Nations General Assembly, 1990). Published as the Ksentini Report, the study stressed the importance of environmental information access (Commission on Human Rights, 1994: 50–53) and participation (Commission on Human Rights, 1994: 53–54). Ksentini's Draft Principles on Human Rights and the Environment were ambitious, providing for information access and dissemination, environmental rights education, public participation, civil society association and justice (Commission on Human Rights, 1994: 76). It is, in my mind, a loss that Ksentini's Draft Principles were never enacted in binding international law (United Nations High Commissioner for Human Rights, 2011: 10; Cramer, 2009: 84; Falkner, 2021: 219). This is reflective of much of the book's story: whilst the propagation of ambitious solidarist concerns is observable, their growth is counterbalanced by state consent, voluntarism and hence states' ability to use their sovereignty to deny ethical progress when they perceive it to surpass their thresholds of tolerance, or to not serve their interests.

With UNCED (the Rio Earth Summit, see United Nations, 1993), however, came "firm evidence" (Elliott, 2004: 7) that environmental protection was firmly on the agenda. Over 170 states attended, plus over 1400 NGOs, with a parallel civil society conference hosting over 30,000 delegates (Elliott, 2004: 17). It should be admitted from the outset that Rio's key outcome, the nonbinding Rio Declaration, was contested. UNCED's Preparatory Committee heard multiple proposals to declare a substantive right to a healthy environment, but consensus was not reached (Shelton, 2006: 133). This is indicative of broader risks that "laggard" states hinder the "leaders" during collective efforts to codify international environmental agreements (Andresen & Agrawala, 2002; Borzel, 2000; Tobin, 2017).

But what Rio *did* achieve was a powerful foregrounding of procedural environmental rights, enabling Aarhus' legitimate subsequent propagation. Rio played a dual role of (a) enshrining the importance of nonstate actors in environmental decision-making practices whilst (b) living out this "spirit of participation" by ensuring record levels of nonstate participation (Falkner, 2021: 177). Whilst we cannot deny Rio's achievements of inducing creation of the UN Commission on Sustainable Development (United Nations, 1993: 459), precipitating near-total state participation in environmental multilateralism and hosting no less than the signature of the United Nations Framework Convention on Climate Change (UNFCCC) and Convention on Biological Diversity (CBD) (United Nations, 1993: 2), it is Rio's focus on procedural propriety that is most propitious for our study. Such focus is evident in Rio's chief outcomes, the Rio Declaration and Agenda 21.

A solidarist turn towards co-creation and procedural propriety surfaces in Rio Declaration Principle 27 (United Nations, 1993: 8), requiring both "states and people" to "cooperate in good faith and in a spirit of partnership" when implementing Rio and developing international law on sustainable development. It is evident in the preamble, seeking "a new and equitable global partnership through the creation of new levels of cooperation among States, key sectors of society and people" (United Nations, 1993: 3). And it is observed in Principle 7 (United Nations, 1993: 4), which requires states to "cooperate in a *spirit of global partnership* to conserve, protect and restore the health and integrity of the Earth's ecosystem" in the context of common but differentiated responsibilities, placing proportionate burdens of responsibility on developed countries "in view of the pressures their societies place on the global environment and of the technologies and financial resources they command". Concerns for procedural propriety emanate from Principle 17 (United Nations, 1993: 6), mandating EIA, whilst transboundary notification of emergencies and disasters (Principle 18) and transboundary EIA (Principle 19) also manifest themselves (United Nations, 1993: 6). Moreover, and noting earlier difficulties concerning codification of substantive environmental rights, Principle 1 (United Nations, 1993: 3; emphasis added) alludes to a substantive right without explicating it fully: "Human beings are at the centre of concerns for sustainable development. They are entitled to a healthy and productive life in harmony with nature".

However, it is Principle 10 that courts most attention in the context of procedural environmental rights, and it is Principle 10 that was codified by Aarhus. It deserves a full quote here (United Nations, 1993: 5; emphasis added):

> Environmental issues are best handled with the participation of *all concerned citizens*, at the relevant level. At the national level, each individual shall have appropriate *access to information* concerning the environment that is held by public authorities, including information on hazardous materials and activities in their communities, and the *opportunity to participate in decision-making processes*. States shall facilitate and encourage *public awareness and participation* by *making information widely available. Effective access to judicial and administrative proceedings*, including redress and remedy, shall be provided.

This provision paved the way for Aarhus' legitimate propagation. Principle 10 was the defining moment at which the world's states came together and "gave their blessing" to the procedural rights of environmental information access, participation and access to justice. Without it, I doubt the Environment for Europe process (see below) would have gained the traction it did.

Of course, far be it from me to suggest that international order eluded the Rio Declaration. The latter did not step too far from the threshold of states' tolerance. Principle 24 deems warfare, for instance, "inherently destructive of sustainable development" and requires states to "respect international law providing protection for the environment in times of armed conflict and cooperate in its further development" (United Nations, 1993: 7; Pavoni & Piselli, 2022). Principle 25 (United Nations, 1993: 7) regards as "interdependent and indivisible" the goals of "peace, development and environmental protection". And with undertones of *Geopolitik*, "a final flourish echoing the UN Charter" (Elliott, 2004: 19) emerges in Principle 26 (United Nations, 1993: 7; emphasis added) and its injunction that "States shall resolve all their environmental disputes peacefully and by appropriate means". Attempts to stabilise international order were still salient and ongoing, by the time UNCED was convened in Rio in 1992. I have no doubt that they were deliberately iterated in the contexts of the post-socialist world's fracture, and of the Gulf and Yugoslav wars, at that time.

The solidarist turn was reinforced by UNCED's second nonbinding outcome, Agenda 21. The latter effectively constituted Rio's implementation plan. Again, let me not offer blind praise. Agenda 21 is full of what Elliott (2004: 20) termed "diplomatic compromises", indicative of states exercising their consent and asserting their sovereign prerogatives in pursuit of their particular interests. Reference, for instance, to economies in transition was all but removed from Agenda 21 given developing states' concerns that "too extensive a recognition of the former Soviet bloc countries could jeopardize their own development needs" (Elliott, 2004: 20). National interests prevailed. Yet solidarist seeds were sown in Agenda 21's first preambular paragraph (United Nations, 1993: 12; emphasis added):

> Humanity stands at a defining moment in history. We are confronted with a perpetuation of disparities between and within nations, a worsening of poverty, hunger, ill health and illiteracy, and the continuing deterioration of the ecosystems on which we depend for our well-being. However, integration of environment and development concerns and greater attention to them will lead to the fulfilment of basic needs, improved living standards for all, better protected and managed ecosystems and a safer, more prosperous future. *No nation can achieve this on its own*; but *together* we can – in a *global partnership* for sustainable development.

Such solidarist rhetoric—of togetherness and global partnership for humankind's benefit as opposed to solely state or industrial benefit—implies the possibility of expanding states' "we feeling" across greater distances. It warrants the deliberate pooling of sovereignty for common ends, presupposing an ethical horizon that encapsulates more than national interests. Sovereignty went nowhere—Agenda 21 was "first and foremost the responsibility of Governments" (United Nations, 1993: 12)—but more was demanded from sovereignty. State-nonstate co-creation was required (United Nations, 1993: 12; emphasis added), as was governance reform for public participation (United Nations, 1993: 96). Agenda 21's participatory concerns arose in contexts of land management (United Nations, 1993: 128), biological diversity (United Nations, 1993: 213), biotechnology (United Nations, 1993: 230), water management (United Nations, 1993: 277, 282, 300) and no less than a freestanding Chapter 23 on the "strengthening" and "genuine involvement of all social groups" in development (United Nations, 1993: 373). There was surely no clearer impetus for Aarhus (United Nations, 1993: 373; emphasis added):

> One of the fundamental prerequisites for the achievement of sustainable development is broad public participation in decision-making...[T]he need for new forms of participation has emerged. This includes the need of individuals, groups and organisations to participate in environmental impact assessment procedures and to *know about* and *participate in decisions*, particularly those which potentially affect the communities in which they live and work.

What we just witnessed, in the account of Rio, and in much of the above context, was aspirational and pathfinding. The context "laid the table" on which Aarhus' binding provisions (see next chapter, Chapter 4) could be propagated. At this juncture, let me offer closing thoughts. Firstly, the persistence of, and insistence upon, state sovereignty is evident. There was clear concern, in the postwar era, for states to foster as nonviolent a coexistence as possible. This was the cardinal aim: to avoid another global war, and to stabilise a framework in which states' enmities could be nonviolently reconciled. There was a sense, in the above context, that more ambitious solidarist concerns would elude multilateralism in the absence of pluralist coexistence. To achieve international order, there *had* to be respect for sovereignty, state consent, non-intervention and diversity in IR. Reaffirmations of these pluralist principles were necessary, in order for states' delegates to be confident enough to take "little steps forward" in more solidarist endeavours. Rather than observing realist stasis, then, I gauge a certain prudence, a cautious incrementalism, a modest hope that voluntary "little steps forward" would in fact *help stabilise* postwar international society, whilst avoiding undue "spillage" of conflicting state interests, and resultant breakdowns into disorder. Solidarist progress was welcomed, but stable pluralist coexistence was non-negotiable.

In fact I argue, secondly, that the pluralist preservation of diversity was a prerequisite for the propagation of solidarist ambition. A push towards human and environmental *justice* would have been infeasible, and diplomatically dangerous, without the pluralist "safety scaffolding" of international *order* to contain it. The context used legal and diplomatic conduits through which a diversity of developed and developing states sought to "play out" their competing interests. The states only consented to the extent that aspirations were palatable for them. Some laws remained "soft"; we witnessed much "declaration". This indicates the persistence of state consent, a lesson resonating today: ethical progress will occur at a pace, intensity and scale determined by states, comfortable in the knowledge that they can withdraw from the diplomatic "pitch" at will.

That said, thirdly, the real achievement of the above context was to have sown the seeds of the cosmopolitan impulse—to have initiated the drivers of human empowerment, "voice" and co-creation—in such a way that the multilateral environment was favourable for Aarhus' legitimate propagation under Environment for Europe (see below). States, in chief, sowed those seeds, and I daresay the strongest were sown at Rio. It is no coincidence that this propagation occurred at the end of the Cold War, with the onset of democratisation. These latter developments set the *geopolitical* context for Aarhus' emergence. Let us now turn to it.

Geopolitical Context

There is a geopolitical element to the context enabling Aarhus' propagation, and it relates to the *Glasnost* (openness) and *Perestroika* (reform) accompanying the Cold War's end. Indeed, the very multilateral framework in which Aarhus emerged was noted (and criticised) for its implication in democratisation. In a source published in the year of Aarhus' signature (European Environmental Bureau, 1998: 6; emphasis added), European Environmental Bureau's (EEB's) then Secretary-General wrote how his organisation "deplores the tendency of Western politicians and officials to consider the "Environment for Europe" process and its products as a learning exercise for Central and Eastern European countries". But such grievance was not universal. This "learning exercise" was ongoing beforehand (see Bellagio and Kiev, below). There is no indication that the USSR and the broader socialist states were compelled to participate: they did so consensually. And I have not excavated, from the archives, either any (a) cynical instrumentalism on the part of those delegates involved in this "learning process" or (b) tones of colonising injustice, indicating a power imbalance against, or a wilful belittling of, those actors engaged in the "learning". The focus was on *mutual* learning: not monologic dictation. There was enthusiasm, between the superpowers, to learn about one another's environmental management approaches.

Gratifying open-mindedness to dialogue and knowledge sharing was evident as the Cold War came to a conclusion. To make this case, the section firstly observes how the democratising dynamics, surrounding the end of the Cold War, offered the geopolitical context for Aarhus' propagation. Secondly, it assesses the legitimising role of dialogue that occurred between actors hailing from both the western liberal democratic and socialist blocs. Thirdly, it addresses the role played by socialist civil society in helping pave the path for Aarhus' emergence. The lesson is that Aarhus would not have emerged as it did, without the power recalibrations accompanying the end of the Cold War, and the consensual willingness of "East" and "West" to "talk more". There is surely a lesson, there, for practitioners today.

Greening Glasnost: Democratisation and Dialogue

An end to bipolarity

The Cold War was characterised, in IR's lingua franca, by bipolarity. Polarity relates to the distribution, "harnessing and coagulation" (Weaver, 2020: 7) of power between states. States, in this regard, constitute power repositories, with power accrued according to states' capabilities, not least in terms of their conventional and nuclear weapons endowments. Bearing in mind our theoretical framework (see Chapter 2), it will come as no surprise that those most notably concerned for the balance of power were associated with realism (Gaddis, 1987; Mearsheimer, 1990; Waltz, 1979). Cold War bipolarity involved two superpowers maintaining the balance of power. At one end of the "scales" was the US and NATO; at the other, the USSR and Warsaw Pact. Both aimed to establish, develop and consolidate their spheres of influence "on, in, and in relation to sovereign territory", projecting their power centrifugally (Weaver, 2020: 7). Such balancing was said to contain direct superpower conflict "by counterbalancing any state that seeks military superiority" (Kegley & Blanton, 2012: 269). If we entertain this, bipolarity fostered a "long peace" in the Cold War (Gaddis, 1987), contingent on superpower survival, nuclear capability equivalence and politico-diplomatic influence. The superpowers, under this logic, were "managers" of a realist system. IR would only undergo significant major change when the balance of power recalibrated accordingly.

Such recalibration *did* occur, at the end of the Cold War. The latter had entailed as much an ideological balance of power as a nuclear one. The

"North Atlantic" constellation subscribed to postwar market and political liberalism. It largely shared concerns for market and political freedom, and for human and commercial autonomy from state economic planning and political intrusion. Such *liberal* (individualist) *democracy* (representative electoral "people power" as opposed to elite autocracy) permitted humans and their associations to make decisions, and take risks, on matters that affected them. Such risk could be taken with less impediment than in autocratic political regimes (within, of course, the confines of democratically enacted law). It goes without saying that such liberalism was not immune to environmental harm. As discussed in the revolutionist section of Chapter 2, emancipation as human "freeing" risks being a high carbon, "kiss-me-quick", throwaway freedom. It is no coincidence that the US witnessed a strong mobilisation of green civil society (Cox, 2006; Depoe et al., 2004).

Meanwhile, the socialist constellation was characterised by centralised, state-led, "command-and-control" politico-economic and environmental planning (Haar, 1992: 492–493). There was less room for humans and their associations to make autonomous decisions, to express dissent, and to express individual and collective interests if incongruent with state socialist objectives. I consciously observe "less room", by the way, as it would be colonising, naïve and entirely inaccurate to claim an absence of societal progressivism under socialism (Weinthal & Watters, 2010; Yanitsky, 2012). But the state socialist approach to government yielded a firm politico-economic grip. Citizenship was harnessed for state ends; the environment and agriculture were collectivised for state socialist purposes (Iordachi & Bauerkämper, 2014). The triumph of central planning over nature was linked to "the meeting of production targets by the national ministries [which] totally ignored the externalities of actions that led to environmental catastrophes" (Haar, 1993: 16). The human individual and nonhuman nature, under socialism, were subordinate to the state in pursuit of the collective "good". Safeguards for transparency, accountability, participation, redress and safe expression of dissent were antithetical to "command-and-control" government. In the USSR, this actually "fanned nationalist sentiment in the republics and increased the longing for sovereignty" from the Kremlin (Haar, 1992: 484). As such, constituent soviet republics' attempted overcoming of Kremlin-dominated state socialism was paired with reaffirmations of self-determination, and some hunger for representative democracy.

Such hunger was articulated not least in environmental contexts. The environmental ills of state socialism, then, induced a renewal of (a more environmentally conscious) statism, hinging on pluralist iterations of the need for sovereignty and diversity in (post)-socialist IR. It is in this regard that we should understand Stec's (2011: 51) idea that Aarhus was conceived "in recognition of the need to establish basic standards of justice" in states characterised by "rapid industrialisation...an authoritarian tradition...[and] a notion of justice based on expediency, aimed at meeting centralised planning objectives, which prevailed over individual needs, wants and rights" (see Hayward, 2002: 246; Jendrośka, 2005: 14; Mason, 2010: 12). There was, without doubt, "a longing of the newly independent states to find a new voice, a new identity, and a new future that would be welcomed among Western democracies" (Zaharchenko, 2009: 29).

Bearing this in mind, I argue that Aarhus' propagation, under UNECE, would have been infeasible without the democratisation accompanying the Cold War's end. The extent of Environment for Europe's ambition, of "East–Wes" dialogue, would have been impossible in the "nuclear stalemate" (Kissinger, 1956: 349) of superpower enmity. It only became possible with the power recalibration and onset of democratisation that accompanied the dissipation of those enmities. The next task is to pay attention to the environmental dialogue that ensued between the superpowers. It paints a curious picture of their tentative convergence *before* the end of Cold War bipolarity. Such dialogue cultivated fertile territory on which Environment for Europe could facilitate the convergence of democracies, (post-)socialist regimes and civil society, in order to propagate the Aarhus Convention.

'East–West' Environmental Dialogue

That the superpowers experienced limited progress on environmental issues, during the Cold War, is no secret (Harman-Stokes, 1995: 94; Krämer, 2012: 97; Yanitsky, 2012: 923). Yet only scant attention was paid to the Bellagio and Kiev Conferences, which were forums for dialogue between the US and USSR. I regard them as having set favourable geopolitical conditions for Aarhus' propagation, given their ability to physically converge—to "bridge" in literal dialogue—delegates from both "East" and "West". Their story needs telling, for the valuable lessons it should offer thinkers and practitioners today.

The Bellagio Conference on US and Soviet Environmental Protection Institutions, held in August 1991, was co-sponsored by the US' Academy of Arts and Sciences and the USSR's Academy of Sciences (Haar, 1992: 481). Delegates of two erstwhile superpower adversaries met to share their knowledge and experience of environmental protection, and to discuss possibilities for environmental cooperation. Like Kiev (below), Bellagio should remind us all of the value of retaining lines of communication and *possibilities* for dialogue between adversaries, however insufferable the diplomatic circumstances. At Bellagio, both superpowers had more to discuss than initially suspected, and this is understandable given that neither the US (Bullard, 1999; Shiva, 1994) nor USSR (Pryde, 1991; Ziegler, 1987) were the most environmentally virtuous. Bellagio hoped for "future coordination between the two nations" (Haar, 1992: 482). The topics addressed were prescient for Aarhus. The conference discussed the role of "local grassroots movements" and local actors' emergence as "passionate political actors" on both sides of the superpower divide (Haar, 1992: 485). The Bellagio Declaration (1991: 499–500) regarded environmental protection as deserving "distinct representation at the highest ministerial or cabinet level of government", and demanded that "citizens and public officials should have appropriate access" to environmental information. The Bellagio Declaration (1991: 500) requested citizens' "right to participate in the government's environmental decision-making process", and that "citizens and groups affected by an environmental decision and responsible government officials should be able to petition a court to interpret and enforce the environmental laws and to overturn actions taken in violation of such laws". Bellagio thus entailed not only commitment, on the part of both the USSR and US, to the high politics of environmental protection, but also to the very procedural rights that would animate Aarhus' propagation under Environment for Europe. The superpowers permitted a solidarist turn towards civil society, towards the procedural propriety with which humans are empowered to influence decision-making processes. There is a rich lesson, somewhere in Bellagio, about the benefit of keeping states' lines of communication open, and about the benefits of converging to discuss mutual ethical progress, despite potential animosities. It is notable that fewer than five months passed after Bellagio, before the red flag was lowered from the Kremlin and replaced with the Russian tricolour (see Zubok, 2022).

Bellagio was followed by the Kiev Conference on Environmental Protection, in November 1992 (Haar, 1993: 12). It was jointly sponsored by Ukrainian, Russian and US institutions (Haar, 1993: 12). Kiev demonstrated collective will, on the part of all delegations, to enhance public participation in environmental matters. Kiev's official report recorded that "ways of mustering public support and citizen involvement in environmental affairs will have to be developed over time" (Haar, 1993: 17). The environment materialised at Kiev as a springboard from which the superpowers discussed democratisation. The report (Haar, 1993: 18), for instance, recalled that even under Soviet rule, "the Green movement was able to attract young people into political action. "Sunshine laws", environmental impact statements, and public access to data are sine qua non for a young democracy". This is bold rhetoric, reflective of the geopolitical recalibrations underway at the end of the Cold War, and indicative of the dissolution of superpower bipolarity that was occurring at that time. It is noteworthy that access to justice (Haar, 1993: 23–25) figured prominently at Kiev, further enriching the political terrain on which Aarhus would soon be propagated.

But what is most prescient for Aarhus' propagation was Kiev's concern for regionalism in the domain of multilateral environmental protection. It is as though Environment for Europe's push towards Aarhus was prophesied. Kiev's report (Haar, 1993: 25) noted that "lawyers should consider…the duties and responsibilities of members of regional organisations…In addition to preventing environmental harm to neighbors outside the region, regional organisations would affirmatively further general international law within and outside the region…As a new concept, it would be a fruitful topic for further discussion among developed and developing countries". Kiev marked, I argue, a crucial yet overlooked contribution to Aarhus' propagation, because it alluded to the benefits of discretely regional environmental instruments between diverse states, in a discrete context of (post-)socialist democratisation. Kiev further strengthened the favourable conditions in which Aarhus could be propagated under Environment for Europe.

Such convergence is easily either overlooked or forgotten when insufferable diplomatic conditions return. Yet the above dialogue contributed to favourable circumstances for Aarhus' propagation, demonstrating the practical feasibility and normative benefit of "bridging" "East" and "West" in dialogue. The cautious hope of Bellagio and Kiev should help erode unhelpful contemporary rhetoric of (a) superpowers "not talking" and (b) democratisation somehow being "thrown" on the (post-)socialist world by a Western unipolar hegemon. The early dialogue of Bellagio and Kiev involved *both* "East" and "West" consenting to embryonic, mutual environmental commitment and, crucially, to human empowerment thereupon. It evinced commitment, on the part of both "East" and "West", to mutual learning. Sovereignty was never called into question. But it was challenged to become more environmentally conscious and "humankind".

Socialist Civil Society

Having considered high politics at superpower level, and its having set favourable geopolitical conditions for Aarhus' propagation, we now turn to the supplementary role of socialist civil society. It is again no secret that calls for environmental transparency and public participation had animated socialist civil society before the Cold War concluded. Calls for reform, "preceding and indeed precipitating the fall of the Berlin Wall" (Sand, 2003: 491), opened possibilities for democratisation (Pedersen, 2008: 99). 1988 witnessed the emergence of the first USSR-wide green NGO, Socio-Ecological Union (Yanitsky, 2012: 925). Also in 1988, Goskompriroda, the USSR's State Committee for Nature Protection, was established (Zaharchenko, 2009: 4). A year afterwards, the USSR adopted the All-Union Law on the Order of Appeal to a Court of Illegal Actions of State Bodies and Officials Infringing the Rights of Citizens. Until this was enacted, "citizens could not go to a court...to challenge public authorities" (Zaharchenko, 2009: 4). Such progress ensued in the context of Chernobyl, which had prompted the Ukrainian Soviet Socialist Republic to issue a 1990 Decree obliging the State Committee for Nature Protection to collect "all relevant environmental information" and "disseminate it to the public through mass media" (Zaharchenko & Goldenman, 2004: 234). Socialist civil society used these forms of law with ardour. In 1992, for instance, Socio-Ecological Union litigated, in the Supreme Arbitration Court of the Russian Federation, against the Russian Federation's decision to waive taxes for a Russo-American oil joint venture (Zaharchenko, 2009: 7). Socio-Ecological Union lost, but this early litigation was "a powerful illustration that the time for NGOs to use legal strategies had arrived...The air at the time was filled with talk about building the rule of law and civil society" (Zaharchenko, 2009: 7).

This ardour, amongst socialist civil society, was also evident at the Meeting on the Protection of the Environment, held between 16 October and 3 November 1989 in Sofia, Bulgaria, under the auspices of the Conference on Security and Cooperation in Europe (CSCE) (1989).

Amongst others present were the USSR and US (CSCE, 1989: ii). UNECE made contributions, alongside UNEP and International Union for Conservation of Nature (IUCN) (CSCE, 1989: 1). Presciently for Aarhus, and necessarily given what happened at the Meeting (see below), the participants "acknowledge[d] the importance of the contribution of persons and organisations dedicated to the protection and improvement of the environment, and [of] allow[ing] them to express their concerns" (CSCE, 1989: 1). The circumstances surrounding this acknowledgement related to the Meeting's interruption by concerns that campaigners from NGO *Ecoglasnost* were being assaulted outside the venue (UNECE, 2000: 23). In response, the Meeting participants reaffirmed, and had noted in the official record (CSCE, 1989: 1), the participating states'

> respect for the right of individuals, groups and organisations concerned with environmental issues to express freely their views, to associate with others, to peacefully assemble, as well as to obtain, publish and distribute information on these issues, without legal and administrative impediments inconsistent with the CSCE provisions. These individuals, groups and organisations have the right to participate in public debates on environmental issues, as well as to establish and maintain direct and independent contacts at national and international level.

Socialist civil society, I argue, thus supplemented the high politics of the environment. Environmentally minded humans—world society actors in the English School lexicon, and literal agents of *Ecoglasnost*—exercised their latent participatory rights to demand reforms to statist decision-making practices. In but one limited historical instance, humans exercised their environmental voice and were "heard" by state delegates, the latter responding by affirming commitment, in the above quote, to human rights of expression, association, assembly, information use, participation and communication. This was one fleeting moment in the history of environmental accountability, when states responded to human iterations of environmental agency. It strengthened the conditions in which Environment for Europe could legitimately propose the Convention. It is a poignant fact that mere days passed between this incident, at the CSCE meeting, and the fall of the Berlin Wall (see Meyer, 2009).

To close this section on geopolitical context, I made the case that Aarhus would not have emerged, as it did, without the end of the Cold War and onset of democratisation. "Nuclear stalemate" is not conducive to convergence and dialogue between superpowers. Insufferable diplomatic conditions require a heavy focus on stabilisation of international order and avoidance of global war. With the dawn of *Glasnost* and *Perestroika*, and the dissipation of bipolarity, came a more fertile territory on which mutual learning could occur between superpowers, not least at Bellagio and Kiev. I fear that the latter instances of convergence were all but forgotten, and so too their lessons for statespersons today.

The fact that the above geopolitical account is almost entirely state-centric should be no shock. States consented to the above dialogue. States voluntarily set aside their differences to discuss the possibility of environmental convergence. The conclusion of the Cold War brought with it renewals of sovereignty, renewals of statism, but they were oriented somewhat more responsibly and democratically (see Weinert, 2007). Humans as world society actors played "second fiddle": an auxiliary role in this geopolitical context. Human agency had an ethically additive effect on statecraft. The lead actors in Aarhus' propagation were states, collectively investing their sovereignty. It is to Environment for Europe, that very multilateral framework in which Aarhus was negotiated, that I now turn.

Environment for Europe

Having discussed the legal, diplomatic and geopolitical contexts setting the conditions for Aarhus' propagation, I now turn to the multilateral Environment for Europe initiative, a partnership of UNECE member states, UN agencies and nonstate actors that "discuss, decide and join efforts in addressing environmental priorities across the 56 countries of the UNECE region" (UNECE, 2023a: online). It will be no surprise, given the above coverage of democratisation, that Environment for Europe is concerned for helping the Eastern Europe, Caucasus and Central Asia (EECCA) region, alongside south-eastern European states, to improve their environmental management practices (UNECE, 2023b: online; see Dannenmaier, 2008: 39). What follows is an examination of the rhetoric of the Environment for Europe ministerial conferences leading to Aarhus' opening for signature at the fourth conference in 1998.

Negotiating Aarhus

The first Environment for Europe conference was held at Dobříš Castle, Czechoslovakia in 1991. A ground-clearing exercise, it set a programme for an "all-European cooperation strategy" (UNECE, 1991: 6) on environmental policy. Dobříš noted that environmental policy should be founded on "sound democratic principles" and "dialogue between governments and *all relevant groups* in society on *national, continental and global scale*" (UNECE, 1991: 2; emphasis added). NGOs were said to be "integral partners" (UNECE, 1991: 2) in such environmental dialogue. The first ministerial conference therefore laid the co-creative foundations for Aarhus' negotiation, given its call for "participation by a well-informed population in the decision-making processes on environmental matters" (UNECE, 1991: 5).

These foundations were strengthened at the second Environment for Europe ministerial conference in Lucerne, Switzerland in 1993. The resultant Lucerne Declaration (UNECE, 1993) scoped Environment for Europe's political trajectory, endorsing international efforts to more justiciably enshrine information access, public participation and access to justice on the international agenda. Solidarism is evident in Lucerne's opening paragraph (UNECE, 1993: 1), "determined to intensify our cooperation in the field of environmental protection in Europe, to assume our responsibility on the global level and to offer our partnership to the other regions of the world". The ministers affirmed their "*cooperation towards convergence*" on environmental policy, representing a "strong integrating force" that would lay "one of the cornerstones for the construction of the *new Europe*" (UNECE, 1993: 1; emphasis added).

This was pioneer rhetoric, imagining a new future—a democratic, convergent future—characterised by stronger dialogue not only between states, but between state and nonstate actors, the latter termed by Lucerne the "informal sectors" (UNECE, 1993: 3). A cosmopolitan impulse of co-creation was generated early on. Most importantly, Lucerne (UNECE, 1993: 7) mandated UNECE's "elaboration of proposals…for legal, regulatory and administrative mechanisms to encourage public participation in environmental decisionmaking". It is therefore evident that calls for practical mobilisation of a binding Principle 10 instrument were articulated from the outset of Environment for Europe: at a time coinciding with Bellagio and Kiev, with the dissipation of Cold War bipolarity, and with the post-Rio flurry of environmental multilateralism.

By the third conference in Sofia, Bulgaria, in 1995, firm action was being taken to operationalise such mandate. The Sofia Declaration (UNECE, 1995a: 7) laid even firmer foundations for Aarhus' propagation, with the attending ministers believing that it was

> essential that, in accordance with Principle 10...States should give the public the opportunity to participate at all levels in decision-making processes relating to the environment...We call upon all countries in the region to ensure that they have a legal framework and effective and appropriate mechanisms to secure public access to environmental information, to facilitate and encourage public participation...and to provide effective public access to judicial and administrative remedies for environmental harm.

Sofia mandated the creation of a legal mechanism that would protect the procedural rights of environmental information access, public participation and access to justice. Sofia's key outcome, UNECE's (1995b) Draft Guidelines on Access to Environmental Information and Public Participation in Environmental Decision-making (Sofia Guidelines), offered a blueprint for Aarhus. The Sofia Guidelines' preamble (UNECE, 1995b: 2) noted that "promotion of public participation requires the transparency and the accountability of public authorities, thus improving their credibility and strengthening the support for their activities". But whilst the Guidelines offered elaborate information (UNECE, 1995b: 3–4) and participation (UNECE, 1995b: 4–5) provisions, their proposed arrangements for "administrative and judicial proceedings" consumed only five lines of the document, in which the public "*should*" be able to access such proceedings, legal guarantees should be "*suitable*", and wide locus standi were "*desirable*" (UNECE, 1995b: 6; emphasis added).

This preference for information access and participatory opportunity, arguably at the expense of justice, was consistent with two trends. Firstly, access to justice only emerged as a well-consolidated third pillar much later in Aarhus' negotiations (see UNECE, 1997b: 10–11, 15–16). Indeed, the title of the draft Convention only included "access to justice" in the tenth and final preparation and negotiation session (UNECE, 1998c: 6). This, I think, indicates states' reluctance to propagate provisions that might incur a significant increase in citizen-initiated litigation. Secondly, the elusiveness of justice reflects broader difficulties operationalising justice provisions internationally, in European jurisdictions (see Poncelet, 2012).

Nevertheless, Sofia was pivotal in refining Environment for Europe's trajectory and contributing to the propagation of the Aarhus Convention. It rallied the practical groundwork necessary. Responding to Sofia, in January 1996, UNECE's Committee on Environmental Policy established an Ad Hoc Working Group (the Working Group) to prepare a draft convention on environmental information access and public participation in environmental decision-making (UNECE, 1996b: 4). State-centric solidarism begins to resonate more acutely at this juncture, despite access to justice figuring neither in the draft title nor Working Group's name. The Working Group's mandate (UNECE, 1996b: 6) is instructive. NGOs involved in Environment for Europe were invited to participate in the drafting, alongside all UNECE member states and the EC. The aim was for the Convention to be prepared and negotiated, by and amongst state and nonstate actors, in enough time for it to open for signature at the fourth Environment for Europe conference in Aarhus, Denmark, in 1998 (Day, 2015: 182; UNECE, 1996b: 6). Sofia, and the subsequent functioning of the Working Group, thus marked a major push, by states, towards a civil society-oriented environmental accountability agreement, designed co-creatively, and thus setting a co-creative precedent for Aarhus' subsequent germination and growth (Jendrośka, 2005: 14; Wates, 2005: 7; see this book, Chapters 4 and 5). This push was, of course, not unanticipated (Hallo, 2011: 60). But it was bold, given the onset of democratisation that had taken hold in the "new Europe".

The Working Group held ten sessions between 1996 and 1998. What is striking is the relative parity of actorhood enjoyed by nonstate actors. Special reference should be made to the impact and influence of the Environmental NGOs Coalition (the Coalition). Environment for Europe permitted, indeed encouraged, an "opening up" of multilateralism to nonstate actors such as the Coalition. The Working Group's first sitting set participatory precedent. State and international organisation delegates were accompanied by representatives of the Coalition, GLOBE Europe, International Council of Environmental Law, Regional Environmental Center, and World Conservation Union (UNECE, 1996c: 1). The Director of UNECE's Environment and Human Settlements Division (UNECE, 1996c: 2) welcomed such an inclusive disposition, expressing "hope that it would strengthen the convention". What is certain is that such state-public dialogue never dissipated during the Convention's negotiation and preparation. It intensified. At the second session, multiple nonstate actors furnished the Working Group with resources for use in the proceedings. These included a "publication on public participation prepared by a non-governmental organisation from the Russian Federation, the report of the pan-European Environmental Citizens Organisations Conference on public participation" and two World Wildlife Fund publications (UNECE, 1996d: 2).

Such relative parity of actorhood continued at the Working Group's third session, when the latter's Chairman "thanked the International Council of Environmental Law...for the support provided...and for the kind offer to assist the Working Group in its further work" (UNECE, 1997f: 3). At this stage, the Coalition gained traction in terms of its influence and impact. The Working Group "agreed to work with" (UNECE, 1997f: 4) a submission by the Coalition, and the latter began to meticulously interrogate, and feed into, the draft's legal wording (UNECE, 1997f: 5–9). It is of course the case that states made abundant interventions (UNECE, 1997f: 13–14, 28, 31)—this was, after all, the negotiation of a new international environmental agreement whose Parties would be consenting states—but they did so with and alongside nonstate actors (UNECE, 1997f: 29–30). By the fourth session, the Coalition's proposals (UNECE, 1997c: 12, 28–33) were being presented "as a *contribution towards deliberations* of the Working Group...*for use* by the drafting committee" (UNECE, 1997c: 28; emphasis added). And by the sixth session, there can be no denial that the Coalition had consolidated itself as a vocal, legitimate negotiating actor in its own right. Account was taken of its legal preferences (UNECE, 1997e: 5–6); several states were recorded as having "supported" the Coalition's proposals (UNECE, 1997e: 6). The Coalition's autonomy is evident in that it (a) "*undertook* to provide wording" for a specific provision (UNECE, 1997e: 6; emphasis added) and (b) submitted a joint proposal with Germany (UNECE, 1997e: 17). This has gravity, vis-à-vis co-creation, as it indicates the Coalition's actor autonomy and joint working practices. This was far from a token gesture of inclusion: state delegations were taking nonstate actors seriously.

The push towards conclusion of Aarhus' preparations and negotiations demonstrated only further intensification of such co-creation. At the Working Group's eighth session, the Italian Minister Plenipotentiary expressed that "his Government was in favour of a *continuously active role* for the non-governmental organisations, not only during the drafting of the convention, but also after its signing" (UNECE, 1997a: 2; emphasis added). This lends credence to the idea that Environment for Europe's "opening up" to civil society set precedent for Aarhus' germination and growth. The Coalition made extensive interventions on legal wording, and its rhetoric became bolder. The Coalition expressed "dismay" at what it deemed certain untransparent negotiating practices (UNECE, 1997a: 3). In the ninth session, the Coalition "strongly deplored" what it deemed a "minimalist" approach to compliance (UNECE, 1998b: 5; see this book, Chapter 5 for further context). By the final sitting, the Coalition had "objected to the dilution" of certain provisions; claimed "disappointment" at others; lamented an "unacceptable level of discretion" afforded to states in the draft; and "*wanted*" the Convention "to ensure the same level of participation for NGOs at meetings of the Parties as had occurred during the negotiations" (UNECE, 1998c: 4; emphasis added). The rhetoric indicates, here, a certain parity of actorhood, benefiting nonstate actors engaged in creating international law, and thereby undermining classical understandings of international law that would regard states as yielding supremacy over law-making (Berman, 2015; Boyle, 2014; Patterson, 2016; Werner, 2016). Such bold, perhaps even belligerent rhetoric, on the part of participating nonstate actors, attempted to render irreversible the co-creative precedent set during Aarhus' negotiations.

Aarhus' negotiations were indeed concluded by the fourth Environment for Europe conference in Aarhus, Denmark between 23 and 25 June 1998 (Brady, 1998: 69). There, Aarhus opened for signature. 52 UNECE member states were represented alongside over 70 IGOs and NGOs (UNECE, 1998d: 4). Kofi Annan, then UN Secretary-General, noted the NGOs' "strong involvement" in the fourth conference and in the Convention's preparations (UNECE, 1998d: 4). The Aarhus Declaration (UNECE, 1998a: 10) regarded the Convention as "provid[ing] recognition for citizens' rights in relation to the environment" and as "a significant step forward both for the environment and for democracy". It (UNECE, 1998a: 10; emphasis added) cast humans and their associations as catalysts of both environmental and democratic health:

> We recognize and support the crucial role played in society by environmental NGOs as an important *channel* for articulating the opinions of the environmentally concerned public. An engaged, critically aware public is essential to a healthy democracy. By helping to empower individual citizens and environmental NGOs to play an active role in environmental policy-making and awareness raising, the Aarhus Convention will promote responsible environmental citizenship and better enable all members of society to fulfil their duty, both individually and in association with others,

> to protect and improve the environment for the benefit of present and future generations.

This appreciation of humans and their associations was acted upon. The Aarhus Declaration (UNECE, 1998a: 10; emphasis added) recorded how the fourth conference held a "*special dialogue* with NGOs", underscoring "their essential role, and our engagement to strengthen lines of communication between governments and NGOs, including in international forums. We recognize the *new role* played by NGOs in this Conference and we greatly appreciate their leadership in organising and taking responsibility for the NGO session and in *actively participating* in Conference preparations".

In a Resolution (UNECE, 1998d: 67) recorded in the official account of the fourth ministerial conference, Aarhus' Signatories "commend[ed] the international organisations and non-governmental organisations, in particular environmental organisations, for their active and constructive participation in the development of the Convention and recommend[ed] that they should be allowed to participate in the same spirit in the Meeting of the Signatories and its activities to the extent possible". This constituted formal recognition, by consenting states, that the co-creation animating Aarhus' propagation would continue during its subsequent germination and growth phases. The following analyses (see Chapters 4 and 5) find scarce contraindications.

Yet there *are* signs of pluralist persistence—of reaffirmations of sovereignty, of eligible states' reluctance to consent—and it would be misleading of me to ignore them. Counterbalancing in favour of state sovereignty was evident. Such reaffirmations serve to delimit the extent of solidarist ethical progress, preventing it from surpassing the threshold of states' tolerance. With this in mind, and having accounted for the negotiations, the next section pays attention to the statist power dynamics underway at the time Aarhus opened for signature and entered into force.

Sovereign Counterbalancing at Aarhus' 'Launch'

Pluralism very much persisted during Aarhus' propagation. State consent was integral to achieving the "humankind" progress during, and beyond, Aarhus' negotiations. Without state consent, without states giving their "blessing" to such partnership with nonstate actors during Aarhus' negotiations, these solidarist endeavours would have fallen on deaf ears.

Indeed, some particularly instructive evidence of the *persistence of*, and *insistence upon*, sovereignty now requires accounting for.

Firstly, neither the US nor Canada participated in Aarhus' preparations and negotiations, despite both being UNECE members (Pallemaerts, 2011: 3). This had a "profound impact" (Jendrośka, 2005: 14) on the Convention's propagation, because both states were endowed with advanced legal frameworks for the discharging of environmental rights. Secondly, whilst the Russian Federation participated in the negotiations, "and in fact many parts of the text were drafted so as to meet the demands of the Russian negotiators", the Russian Federation ultimately refused to sign Aarhus (Wates, 2011: 399–400). In fact, the Russian delegation intended to sign, but "at the last minute" was directed by the Kremlin not to (Zaharchenko, 2009: 52). This was because, at that time, Russia's defence ministry and Federal Security Service (FSB) were "strongly opposed" to Aarhus (Zaharchenko, 2009: 25). A key determinant will have been the action taken by Alexander Nikitin, awarded the 1997 Goldman Environmental Prize (Goldman Environmental Foundation, 2023: online). A former naval officer, Nikitin was imprisoned by the Russian Federation for treason and divulging state secrets, after being "stung" by law enforcement whilst attempting to draft a report entitled "The Russian Northern Fleet: Sources of Radioactive Contamination". Nikitin was acquitted in 1999, after his case was taken to the Russian Federation's Supreme Court (Goldman Environmental Foundation, 2023). The case will have unnerved the Russian executive and security architecture, causing its Aarhus delegation to renege on its intentions.

I employ these vignettes because they indicate that on one hand, states that *could* consent to entering the diplomatic "pitch" exercised their right not to. Meanwhile, on the other hand, a state that ultimately exercised its sovereign prerogative to leave the "pitch", and refuse to sign Aarhus, nonetheless yielded influence on the drafting of the Convention text, thus contributing to an alteration of environmental legal, regulatory and governance practices inside other states' domestic jurisdictions. The US, Canada and the Russian Federation pursued their interests through an international diplomacy of abstinence. Such voluntarism strikes a chord with Belarus' later withdrawal from the Convention (see Chapter 5).

Similarly, some states, upon signing Aarhus, exercised their right under international law (Vienna Convention on the Law of Treaties, 1969, Article 2.1.d) to unilaterally express reservation about Aarhus' applicability to them. Denmark stated (UNECE, 1998d: 70) that the autonomous territories of Faroe Islands and Greenland were "scarcely populated and far less diverse" than the "European countries with relatively large populations and corresponding administrative and social structures", and thus "full implementation of the Convention in these areas [Faroe Islands and Greenland] may imply needless and inadequate bureaucratisation". Danish signature did "not necessarily mean...that Danish ratification will in due course include the Faroe Islands and Greenland" (UNECE, 1998d: 70). I deem this an attempted mitigation of the pursuit, by those favouring Faroese and Greenlandic independence, of the territories' further autonomy. Danish sovereignty over natural resources, ocean governance and energy was scrutinised, during and after the Cold War (Dragsdahl, 2001; Floistad, 1988; Loukacheva, 2007; Nuttall, 2008; Orvik, 1984). I suspect that there was unease, amongst opponents to the territories' autonomy, with the further empowerment of Faroese and Greenlandic populations to more effectively influence decision-making practices that might affect them and their environments.

Alongside Denmark, the UK exercised reservation, on signature, deeming "an aspiration" (UNECE, 1998d: 72) Aarhus' recognition of a nonbinding substantive environmental right (Convention on Access to Information, Public Participation in Decision-making and Access to Justice in Environmental Matters, 1998, seventh Preambular paragraph, Article 1; see this book, Chapter 1). The UK limited the purview of its legal guarantees to Aarhus' operative procedural rights to information access, public participation and access to justice (UNECE, 1998d: 72; see this book, Chapter 4). The UK used its sovereign discretion to make unequivocally clear its legal position vis-à-vis the absence of legally binding substantive rights. As such, the traditional capacity, under international law, for states to exercise reservation about the applicability of legal provisions to them, counterbalanced against the relative parity of actorhood enjoyed by nonstate actors.

The persistence of, and insistence upon, sovereignty is further evident when we subject Aarhus' (a) original signatories and (b) ratifying states to scrutiny. I will now study the composition of the original signatories, at the fourth Environment for Europe conference, before making sense of the sixteen ratifying states that triggered Aarhus' entrance into force (see Convention on Access to Information, Public Participation in Decision-making and Access to Justice in Environmental Matters, 1998,

Article 20.1). Using UN (2020: 165) classifications, the following observations are made about the 35[1] original signatories, notwithstanding the EC (UNECE, 1998d: 69). Two of the 35 original signatories, Liechtenstein and Monaco (circa 6%), did not ratify. Enigmatically, the Aarhus Convention Clearinghouse (2023: online; emphasis added) states that Liechtenstein did not ratify "due to *limited human resources*". This is curious, given that the same source (Aarhus Clearinghouse for Environmental Democracy, 2023: online) identifies Liechtenstein as "Party to 28 conventions and protocols in the area of climate, the Alps, biological diversity, air pollution control, chemicals and water pollution control". Meanwhile, at the time of writing, the Aarhus Convention Clearinghouse contains no resources on Monaco. Monaco's reasons for not ratifying are not readily apparent. What I can safely remark is that Monaco is no stranger to UN international law, given that it signed, ratified, acceded to, approved or otherwise accepted around 60 UN treaties since joining the UN in 1993, with around one-third being environmental agreements; more than half of the latter were ratified, approved, accepted or acceded to (Permanent Mission of Monaco to the United Nations, 2023: online). Both Liechtenstein and Monaco are principalities, and at the time of writing, Liechtenstein ranks first in the world by real gross domestic product per capita, and Monaco third (Central Intelligence Agency, 2023a, b, c).

Things become more interesting when we examine the discrepancies between Aarhus' original "leader" signatories and the "trigger" states whose 16 ratifications met the quorum for Aarhus' entrance into force (Convention on Access to Information, Public Participation in Decision-making and Access to Justice in Environmental Matters, 1998, Article 20.1). Of the 35 original signatories, over 80% are developed economies[2] whilst around 20% are economies in transition.[3] Just over one-third[4] of the original signatories had some historical roots in the Warsaw Pact, whilst just under 70% of Aarhus' original signatories are now NATO members.[5] It is worth contrasting this data with the 16 ratifying "trigger" states, because the tables appeared to have turned: of them, only 25%[6] are developed economies; 75%[7] are economies in transition; over 80%[8] had some historical roots in the Warsaw Pact; and just under 40%[9] are now NATO members. Whilst the original "leaders" were largely "NATO-friendly" developed economies, the ratifying "triggers" were largely post-socialist economies in transition. This indicates the possibility of a democratising dynamic underway, between Aarhus' signature and entrance into force. Conditionality (Elliott, 2004: 180) may also help make sense of this: post-socialist states will have been encouraged, through political, diplomatic and trade channels, to demonstrate "good governance" commitments (Nanda, 2006) in return for investment, as the new millennium dawned. Another factor is the deliberate socialisation of post-socialist states into more (market and politically) liberal, global, environmentally aware post-Cold War arrangements (Weaver, 2015: 46–58; see Compagnon et al., 2012: 240).

As such, it is possible to observe a distinct statism during Aarhus' launch. It is clear that the push towards environmental accountability hinged solidly on states' willing consent. The chief lesson to be drawn is that moral cosmopolitan efforts to codify "humankind" governance arrangements, in international law, will be tempered by states, to ensure that those arrangements do not surpass their thresholds of tolerance. Solidarist ethical progress will only be permitted by states to go as far as they deem appropriate to their interests. States can, and do, leave the diplomatic "pitch" when solidarist endeavours no longer serve their interests, or when they deem those endeavours to have "gone too far". This is apparent in the Russian Federation's "last minute" refusal to *sign*; in US and Canadian refusal to *negotiate*; in Danish and UK iterations of sovereign reservation; in Liechtensteiner and Monacan refusals to *ratify*; and in the curious—and inherently statist—discrepancies between the "leader" Signatories and ratifying "trigge" states. Whilst I do not deny the relative parity of actorhood enjoyed by nonstate actors during Aarhus' propagation, and whilst it indicates a customary warmth towards civil society that was garnered in the postwar era and never lost in the Aarhus regime, it has to be admitted that states, and only states, mutually invested their sovereignty in such "humankind" multilateralism.

Conclusion

The purpose of this chapter was to assess the international context that set favourable conditions for, and contributed to, the propagation of the Aarhus Convention. It examined the corpus of international legal and diplomatic context, before turning to the geopolitical context of the Cold War's end and onset of democratisation. Afterwards, the chapter accounted for Environment for Europe, the UNECE initiative under which state and nonstate actors engaged in a co-creative process that resulted in the Aarhus Convention. Before closing, attention was paid to the residual statism evident during Aarhus' opening for signature and entrance into force.

Aarhus' propagation cannot be separated from these state-oriented contexts. I claimed, at the start of the chapter, that Aarhus did not "spring" its ambitions upon environmental multilateralism without warning. I hope to have demonstrated that whilst it is a bold pioneer, driven by a cosmopolitan impulse of human empowerment, voice and co-creation, Aarhus represents the accumulation of, and elaboration upon, a wealth of international effort, by states in the postwar era, to converge upon the pressing issues of human rights and the environment.

My chief finding is that Aarhus' propagation was only feasible to the extent that its cosmopolitan impulse was contained and legitimised by states. Any sense of human and environmental justice would have failed to make the agenda, had it not been compatible with, and palatable for, the cardinal goal of postwar international society: to avoid global war, and to foster as nonviolent and stable an international coexistence as possible. In this regard, I would go so far as to argue that pluralist guarantees—for states to continue to reaffirm and iterate their sovereignty—constituted a prerequisite for the development of ethically stricter, more other-oriented solidarist ambition. I argue that the more ethically exacting pursuit of human and environmental justice is simply infeasible without a "safety scaffolding" of nonviolent international order between states. Take the scaffolding away, remove the order, and you reduce to rubble the likelihood of effectively pursuing—let alone achieving—the more ambitious objectives of human and environmental rights-claiming.

The reaffirmations of sovereignty had to be reinforced in the above contexts, in order for states to be comfortable and confident taking incremental, "little steps forward" in their solidarist endeavours. After all, we saw—as late as Rio—that concerns for sustainable development, human empowerment and environmental protection ran parallel with the fundamental priority of avoiding violence between states. The salient priority, throughout the above contexts, was the assurance of nonviolence, alongside preservation of human dignity, in order to avoid the tragedies of *in*dignity accompanying global war. In this regard, the pluralist pursuit of international *order* by states was a precondition for the solidarist pursuit of *justice*.

As such, a delicate balancing act was evident. States sought (a) aspirational, often non-justiciable solidarist objectives whilst counterbalancing them against (b) their national interests of security, development, and extractive growth. Some of the earlier aspiration, for instance, paired belligerent, extractive and colonising goals with considerations for living conditions, environmental health and industrial hygiene.

Such balancing was also evident in states' unilateral abstentions and reservations. The competing priorities of stabilising international order, preserving national interests, pursuing environmentally intrusive development objectives, and envisaging a political trajectory of "humankindness" resulted in tensions, inconsistencies and compromises between solidarist rationale and pluralist reality. This finding echoes Jackson's (2003: 417) conclusion that international law constitutes a response "to the facts of diversity and imperfection" that fits "the political facts of the world" and avoids requiring actions that go "beyond what most people and countries are prepared and equipped to deliver".

Of course, I do not deny the strength of the cosmopolitan impulse sown. The NGO Coalition, for example, enjoyed such a rich parity of actorhood, during Aarhus' propagation, that it ultimately managed to enjoy some very limited degree of functional hybridity. The close partnership working between state and nonstate actors, and the inclusive disposition towards the latter, set a co-creative precedent that never dissipated after Aarhus' entrance into force. But what I must reiterate, following Jackson's understanding of pluralism and solidarism, is that such civil society actorhood is by no means independent of the orthodox power frameworks of international society. Rather, it is "made possible by and is underwritten by" (Jackson, 2003: 417) states' willing and voluntary consent within already existing international society arrangements. The gratifying capacity granted to civil society occurred because consenting states "gave their blessing" to it, deeming it in their interests, and in the interests of their populations and environments, to do so.

Implications are yielded for solidarisation. I observed counterbalancing between the sowing of the cosmopolitan impulse and the reaffirmation of sovereignty in a world prevailed upon by states. I argue that a vital safeguard for Aarhus' propagation was the preservation of states' capacity to only take solidarist commitments as far as they benefited *the states themselves*. States' ability to leave the diplomatic "pitch", and to iterate their sovereign reservations, serves to reassure state delegates that sovereignty retains primacy as the definitive currency of IR. Such counterbalancing in favour of sovereignty emboldens states to consent to a cautious, incremental cosmopolitanism, comfortable in the knowledge that they can retract at will.

Through the lens of solidarisation, solidarist endeavours can thus "never be devoid of a pluralist core" (Ahrens, 2018: 97). Their transformative potential is contained by, and confined within, existing power frameworks. Pluralism, I repeat, is a precondition for solidarism. The world's diversity was, and remains, such that any rights norms that *are* successfully enacted "must be based on standards of conduct that statespeople everywhere can acknowledge regardless of their own cultural backgrounds" (Jackson, 2003: 410). If state support eludes solidarist endeavours, they will remain consigned to the realm of aspiration, and will not bear fruit. In sum, solidarist progress will "hardly be radical" and must "be incorporated in previously existing structures in order to be perceived as legitimate" (Ahrens, 2019: 268).

Notes

1. The Aarhus Convention's 35 original signatories, notwithstanding the European Community, were: Albania, Armenia, Austria, Belgium, Bulgaria, Croatia, Cyprus, Czech Republic, Denmark, Estonia, Finland, France, Georgia, Greece, Iceland, Ireland, Italy, Kazakhstan, Latvia, Liechtenstein, Lithuania, Luxembourg, Monaco, Netherlands, Norway, Poland, Portugal, Moldova, Romania, Slovenia, Spain, Sweden, Switzerland, Ukraine and the UK.
2. The 29 (circa 83%) original Aarhus Convention signatories that are now considered developed economies, using United Nations (2020) classifications, were: Austria, Belgium, Bulgaria, Croatia, Cyprus, Czech Republic, Denmark, Estonia, Finland, France, Greece, Iceland, Ireland, Italy, Latvia, Liechtenstein, Lithuania, Luxembourg, Monaco, Netherlands, Norway, Poland, Portugal, Romania, Slovenia, Spain, Sweden, Switzerland and the UK.
3. The 7 (circa 20%) original Aarhus Convention signatories that are now considered economies in transition, using United Nations (2020) classifications, were: Albania, Armenia, Georgia, Kazakhstan, Moldova, North Macedonia and Ukraine.
4. The 13 (circa 37%) original Aarhus Convention signatories that had some historical roots in the Warsaw Pact were: Albania, Armenia, Bulgaria, Czech Republic, Estonia, Georgia, Kazakhstan, Latvia, Lithuania, Moldova, Poland, Romania and Ukraine.
5. The 24 (circa 69%) original Aarhus Convention signatories that are now NATO members were: Albania, Belgium, Bulgaria, Croatia, Czech Republic, Denmark, Estonia, France, Finland, Greece, Iceland, Italy, Latvia,

Lithuania, Luxembourg, North Macedonia, Netherlands, Norway, Poland, Portugal, Romania, Slovenia, Spain and the UK.

6. Of the 16 quorum states triggering Aarhus' entrance into force, the 4 that would now be classed as developed economies are: Denmark, Hungary, Italy and Romania.
7. Of the 16 quorum states, the 12 economies in transition were: Albania, Armenia, Azerbaijan, Belarus, Georgia, Kazakhstan, Kyrgyzstan, Moldova, North Macedonia, Tajikistan, Turkmenistan and Ukraine.
8. Of the 16 quorum states, the 13 with some historical roots in the Warsaw Pact were: Albania, Armenia, Azerbaijan, Belarus, Georgia, Hungary, Kazakhstan, Kyrgyzstan, Moldova, Romania, Tajikistan, Turkmenistan and Ukraine.
9. Of the 16 quorum states, the 6 that are now NATO members are: Albania, Denmark, Hungary, Italy, North Macedonia and Romania.

Bibliography

Ahrens, B. (2018). *AmbigEUity: the EU and the solidarisation of international society.* (PhD thesis), Eberhard Karls Universität Tübingen.

Ahrens, B. (2019). The European Union between solidarist change and pluralist re-enactment, In T. Brems Knudsen & C. Navari (Eds.), *International Organisation in the Anarchical Society: the Institutional Structure of World Order.* (pp. 265–292). Palgrave Macmillan

Aarhus Clearinghouse for Environmental Democracy. (2023). *International environmental agreements ratified and signed by Liechtenstein.* Available at: https://aarhusclearinghouse.unece.org/resources/international-environmental-agreements-ratified-and-signed-liechtenstein?individual=1&field_resource_geography_value=1000050. Accessed 12 April 2023.

Andresen, S., & Agrawala, S. (2002). Leaders, pushers and laggards in the making of the climate regime. *Global Environmental Change, 12*(1), 41–51.

Barritt, E. (2020). *The foundations of the Aarhus convention: Environmental democracy, rights and stewardship.* Hart.

Bellagio Declaration on the Environment. (1991). *Boston College Environmental Affairs Law Review, 19*(3), 499–500.

Berman, P. (2015). Nonstate law making through the lens of global legal pluralism. In M. Helfand (Ed.), *Negotiating state and nonstate law: The challenge of global and local legal pluralism* (pp. 15–40). Cambridge University Press.

Birnie, P. W., Boyle, A., & Redgwell, C. (2009). *International law and the environment.* Oxford University Press.

Blundell, D. (2015). Access to and collection of environmental information. In C. Banner (Ed.), *The Aarhus convention: A guide for UK lawyers* (pp. 99–120). Hart.
Borzel, T. (2000). Why there is no 'southern problem': On environmental leaders and laggards in the European Union. *Journal of European Public Policy, 7*(1), 141–162.
Boyle, A. (2014). Soft law in international law-making. In M. Evans (Ed.), *International law* (pp. 118–136).
Brady, K. (1998). New convention on access to information and public participation in environmental matters. *Environmental Policy and Law, 28*(2), 69–75.
Braig, K. F., Kutepova, N., & Vouleli, V. (2022). Playing Second fiddle to the Aarhus convention: Why the ECtHR can and should go further. *Journal for European Environmental and Planning Law, 19*, 74–102.
Briant, J. (2017). Unheard voices: A critical discourse analysis of the Millennium Development Goals' evolution into the Sustainable Development Goals. *Third World Quarterly, 38*(1), 16–41.
Brown, C., & Ainley, K. (2009). *Understanding international relations*. Palgrave Macmillan.
Bullard, R. (1999). Environmental justice challenges at home and abroad. In N. Low (Ed.), *Global ethics and environment* (pp. 33–46). Routledge.
Calvocoressi, P. (2013). *World politics since 1945*. Routledge.
Carson, R. (1962). *[2000] Silent spring*. Penguin.
Central Intelligence Agency. (2023a). *The world factbook: Liechtenstein*. Available at: https://www.cia.gov/the-world-factbook/countries/liechtenstein/#government. Accessed 14 April 2023.
Central Intelligence Agency. (2023b). *The world factbook: Monaco*. Available at: https://www.cia.gov/the-world-factbook/countries/monaco/#government. Accessed 14 April 2023.
Central Intelligence Agency. (2023c). *The world factbook: Real GDP per capita*. Available at: https://www.cia.gov/the-world-factbook/field/real-gdp-per-capita/country-comparison/. Accessed 14 April 2023.
Charter of the United Nations and Statute of the International Court of Justice. (1945). Available at: https://treaties.un.org/doc/publication/ctc/uncharter.pdf. Accessed 15 April 2023.
Coalition for the International Criminal Court. (2023). *Our story*. Available at: https://www.coalitionfortheicc.org/about/our-story. Accessed 15 April 2023.
Commission on Human Rights. (1994). *Human rights and the environment: Final report prepared by Mrs. Fatma Zohra Ksentini, Special Rapporteur*. E/CN.4/Sub.2/1994/9. Available at: https://digitallibrary.un.org/record/226681?ln=en. Accessed 25 April 2023.

Compagnon, D., Chan, S., & Mert, A. (2012). The Changing role of the state. In F. Biermann & P. Pattberg (Eds.), *Global environmental governance reconsidered* (pp. 237–263). MIT Press.

Conference on Security and Co-operation in Europe. (1989). *Report on conclusions and recommendations of the meeting on the protection of the environment of the Conference on Security and Co-operation in Europe*. Available at: https://www.osce.org/files/f/documents/6/9/14075.pdf. Accessed 13 April 2023.

Convention for the Protection of Human Rights and Fundamental Freedoms. (1950). Available at: https://www.echr.coe.int/Documents/Convention_ENG.pdf. Accessed 16 April 2023.

Convention on Access to Information, Public Participation in Decision-making and Access to Justice in Environmental Matters. (1998). ECE/CEP/43. Available at: https://unece.org/DAM/env/pp/documents/cep43e.pdf. Accessed 9 March 2023.

Convention on Environmental Impact Assessment in a Transboundary Context. (1991). *United Nations Treaty Series, 1989*. Available at: https://treaties.un.org/doc/Treaties/1991/02/19910225%2008-29%20PM/Ch_XXVII_04p.pdf. Accessed 21 April 2023.

Convention on Long-Range Transboundary Air Pollution. (1979). *United Nations Treaty Series, 1302*. Available at: https://treaties.un.org/doc/Treaties/1979/11/19791113%2004-16%20PM/Ch_XXVII_01p.pdf. Accessed 21 April 2023.

Convention on the International Right of Correction. (1953). *United Nations Treaty Series, 435* (pp. 191–236). Available at: https://treaties.un.org/doc/Publication/UNTS/Volume%20435/v435.pdf. Accessed 15 April 2023.

Council of Europe. (2023). *Etat des signatures et ratifications du traité 205*. Available at: https://www.coe.int/fr/web/conventions/full-list?module=signatures-by-treaty&treatynum=205. Accessed 16 April 2023.

Council of Europe Convention on Access to Official Documents. (2008). Available at: https://rm.coe.int/access-to-official-documents/16809f5c1a. Accessed 16 April 2023.

Cox, R. (2006). *Environmental communication and the public sphere*. SAGE.

Cramer, B. W. (2009). The human right to information, the environment and information about the environment: From the Universal Declaration to the Aarhus Convention. *Communication Law and Policy, 14*, 73–103.

Curtis, D., & Taylor, P. (2017). The United Nations. In J. Baylis, S. Smith, & P. Owens (Eds.), *The globalisation of world politics: An introduction to international relations* (pp. 331–346). Oxford University Press.

Dannenmaier, E. (2008). A European Commitment to environmental citizenship: Article 3.7 of the Aarhus Convention and public participation in international forums. *Yearbook of International Environmental Law, 18*(1), 32–64.

Darpö, J. (2018). Principle 10 and access to justice. In L. Krämer, & E. Orlando (Eds.), *Elgar encyclopedia of Environmental Law—Principles of Environmental Law* (pp. 379–402).

Day, C. (2015). The Aarhus Convention and NGOs. In C. Banner (Ed.), *The Aarhus Convention: A guide for UK lawyers* (pp. 181–198). Hart.

Depoe, S. P., Delicath, J. W., & Elsenbeer, M. A. (Eds.). (2004). *Communication and public participation in environmental decisionmaking*. SUNY Press.

Dragsdahl, J. (2001). The Danish Dilemma. *Bulletin of the Atomic Scientists, 57*(5), 45–51.

Ebbesson, J. (2011). A modest contribution to environmental democracy and justice in transboundary contexts: The combined impact of the Espoo Convention and Aarhus Convention. *Review of European Community and International Environmental Law, 20*(3), 248–257.

Eicke, T. (2022). Climate change and the Convention: Beyond admissibility. *European Convention on Human Rights Law Review, 3*, 8–16.

Elliott, L. (2004). *The global politics of the environment*. Palgrave Macmillan.

European Environmental Bureau. (1998). *Doors to democracy: Current trends and practices in public participation in environmental decisionmaking in Western Europe*. Available at: https://terracentro.org/?wpfb_dl=36. Accessed 12 April 2023.

Falkner, R. (2021). *Environmentalism and global international society*. Cambridge University Press.

Fasoli, E., & McGlone, A. (2018). The Non-Compliance mechanism Under the Aarhus Convention as 'Soft' enforcement of International Environmental Law: Not so soft after all! *Netherlands International Law Review, 65*, 27–53.

Floistad, B. (1988). Fisheries, Greenland and the European Community: Formal independence, economic dependence. *International Challenges, 8*(4), 6–10.

Gaddis, J. (1987). *The long peace: Inquiries into the history of the Cold War*. Oxford University Press.

Gibboney, C. N. (1949). The United Nations scientific conference for the conservation and utilisation of resources. *Science, 110*(2869), 675–678.

Goldman Environmental Foundation. (2023). *1997 Goldman Prize Winner: Alexander Nikitin*. Available at: https://www.goldmanprize.org/recipient/alexander-nikitin/. Accessed 12 April 2023.

Haar, C. (1992). Foreword: a report on the Bellagio Conference on U.S.-U.S.S.R. environmental protection institutions. *Boston College Environmental Affairs Law Review, 19*(3), 481–498.

Haar, C. (1993). Project report: Kiev Conference on Environmental Protection. *Bulletin of the American Academy of Arts and Sciences, 47*(2), 12–31.

Hallo, R. (2011). Access to environmental information: The reciprocal influences of EU law and the Aarhus Convention. In M. Pallemaerts (Ed.), *The Aarhus Convention at ten: Interactions and tensions between conventional international law and EU environmental law* (pp. 55–66). Europa Law.

Hardin, G. (1968). The tragedy of the commons. *Science, 162*(3859), 1243–1248.

Harman-Stokes, K. M. (1995). Community right-to-know in the newly independent states of the former Soviet Union: Ending the culture of secrecy surrounding the environmental crisis. *Virginia Environmental Law Journal, 15*, 77–138.

Hayward, T. (2002). Environmental rights as democratic rights. In B. Minteer, & B. Pepperman Taylor (Eds.), *Democracy and the claims of nature* (pp. 237–256). Rowman and Littlefield.

Heywood, A. (2014). *Global politics*. Palgrave Macmillan.

HM Parliament. (2004). *Joint committee on human rights: The international covenant on economic, social and cultural rights*. HL 183, HC 1188. Available at: https://publications.parliament.uk/pa/jt200304/jtselect/jtrights/183/183.pdf. Accessed 27 April 2023.

Hurrell, A. (2007). *On global order*. Oxford University Press.

International Covenant on Civil and Political Rights. A/RES/2200A(XXI). Available at: https://treaties.un.org/doc/treaties/1976/03/19760323%2006-17%20am/ch_iv_04.pdf. Accessed 15 April 2023.

International Covenant on Economic, Social and Cultural Rights. A/RES/2200A(XXI). Available at: https://treaties.un.org/doc/treaties/1976/01/19760103%2009-57%20pm/ch_iv_03.pdf. Accessed 15 April 2023.

Iordachi, C., & Bauerkämper, A. (Eds.). (2014). *The collectivisation of agriculture in Communist Eastern Europe: Comparison and entanglements*. Central European University Press.

Jackson, R. (2003). *The global covenant: Human conduct in a world of states*. Oxford University Press.

Jendrośka, J. (2005). Aarhus Convention and Community Law: The interplay. *Journal for European Environmental and Planning Law, 1*, 12–21.

Kegley, C., & Blanton, S. (2012). *World politics: Trend and transformation*. Wadsworth.

Kissinger, H. (1956). Force and nuclear diplomacy in the nuclear age. *Foreign Affairs, 34*(3), 349–366.

Kiwanuka, R. (1988). Developing rights: The UN Declaration on the Right to Development. *Netherlands International Law Review, 35*(3), 257–272.

Krämer, L. (2012). Transnational access to environmental information. *Transnational Environmental Law, 1*(1), 95–104.

Linklater, A. (2002). Cosmopolitan political communities in international relations. *International Relations, 16*(1), 135–150.

Lloyd, L. (1995). The League of Nations and the settlement of disputes. *World Affairs, 157*(4), 160–174.
Loukacheva, N. (2007). *The Arctic promise: Legal and political autonomy of Greenland and Nunavut*. University of Toronto Press.
Lundestad, G. (2017). *International Relations since 1945: East, West, North, South*. SAGE.
Mason, M. (2010). Information disclosure and environmental rights: The Aarhus Convention. *Global Environmental Politics, 10*(3), 10–31.
Mearsheimer, J. (1990). Back to the future: Instability in Europe after the Cold War. *International Security, 15*(1), 5–56.
Meyer, M. (2009). *The year that changed the world: The untold story behind the fall of the Berlin Wall*. Scribner.
Montreal Protocol on Substances that Deplete the Ozone Layer. (1987). *United Nations Treaty Series, 1522* (pp. 28–111).
Nanda, V. (2006). The 'good governance' concept revisited. *Annals of the American Academy of Political and Social Science, 603*(1), 269–283.
Nuttall, M. (2008). Self-rule in Greenland: Towards the world's first independent Inuit state? *Indigenous Affairs, 3–4*, 64–70.
Office of the High Commissioner for Human Rights. (2018). *Universal Declaration of Human Rights at 70: Article 28*. Available at: https://www.ohchr.org/en/press-releases/2018/12/universal-declaration-human-rights-70-30-articles-30-articles-article-28#:~:text=Article%2028%3A%20Right%20to%20a%20Free%20and%20Fair%20World&text=Article%2028%20says%2C%20in%20its,Declaration%20can%20be%20fully%20realised.%22. Accessed 15 April 2023.
Okowa, P. (1997). Procedural obligations in international environmental agreements. *British Yearbook of International Law, 67*(1), 275–336.
Orvik, N. (1984). Greenland: The politics of a new Northern nation. *International Journal, 39*, 932–961.
Pallemaerts, M. (2011). Introduction. In M. Pallemaerts (Ed.), *The Aarhus Convention at ten: Interactions and tensions between conventional international law and EU environmental law* (pp. 1–16). Europa Law.
Patterson, D. (2016). Transnational lawmaking. In C. Brölmann & Y. Radi (Eds.), *Research handbook on the theory and practice of international lawmaking* (pp. 56–65). Edward Elgar.
Pavoni, R., & Piselli, D. (2022). *Armed conflicts and the environment: An assessment of Principle 24 of the Rio Declaration Thirty Years On*. Available at: https://papers.ssrn.com/sol3/papers.cfm?abstract_id=4071106. Accessed 25 April 2023.
Pedersen, O. W. (2008). European environmental human rights and environmental rights: A long time coming? *Georgetown International Environmental Law Review, 21*, 73–111.

Permanent Mission of Monaco to the United Nations. (2023). *Signature, ratification, accession, approval and acceptance of United Nations Treaties since the Principality' s admission to the Organisation on 28 May 1993.* Available at: https://mission-un-ny.gouv.mc/Monaco-and-the-UN/Treaties-of-the-United-Nations. Accessed 12 April 2023.

Peters, B. (2018). Unpacking the diversity of procedural environmental rights: The European Convention on Human Rights and the Aarhus Convention. *Journal of Environmental Law, 30*, 1–27.

Poncelet, C. (2012). Access to justice in environmental matters: Does the European Union comply with its obligations? *Journal of Environmental Law, 24*(2), 287–309.

Protocol to the Convention for the Protection of Human Rights and Fundamental Freedoms. (1952). Available at: https://www.echr.coe.int/Documents/Convention_ENG.pdf. Accessed 16 April 2023.

Pryde, P. (1991). *Environmental management in the Soviet Union.* Cambridge University Press.

Sand, P. H. (2003). Information disclosure as an instrument of environmental governance. *Heidelberg Journal of International Law, 63*, 487–502.

Shelton, D. (2006). Human rights and the environment: What specific environmental rights have been recognised? *Denver Journal of International Law and Policy, 35*(1), 129–171.

Shiva, V. (1994). Conflicts of global ecology: Environmental activism in a period of global reach. *Alternatives, 19*(2), 195–207.

Stec, S. (2011). EU enlargement, neighbourhood policy and environmental democracy. In M. Pallemaerts (Ed.), *The Aarhus Convention at ten: Interactions and tensions between conventional international law and EU environmental law* (pp. 35–54). Europa Law.

Tobin, P. (2017). Leaders and laggards: Climate policy ambition in developed states. *Global Environmental Politics, 17*(4), 28–47.

United Nations. (1972). *Report of the United Nations Conference on the Human Environment.* A/CONF./48/14/Rev.1. Available at: https://documents-dds-ny.un.org/doc/UNDOC/GEN/NL7/300/05/IMG/NL730005.pdf?OpenElement. Accessed 18 April 2023.

United Nations. (1993). *Report of the United Nations Conference on Environment and Development, Rio de Janeiro, 3–14 June 1992.* A/CONF.151/26/Rev.1. Available at: https://documents-dds-ny.un.org/doc/UNDOC/GEN/N92/836/55/PDF/N9283655.pdf?OpenElement. Accessed 14 April 2023.

United Nations. (2020). *World economic situation and prospects.* Available at: https://www.un.org/development/desa/dpad/wp-content/uploads/sites/45/WESP2020_Annex.pdf. Accessed 9 March 2023.

United Nations. (2023). *United Nations Charter.* Available at: https://www.un.org/en/about-us/un-charter. Accessed 15 April 2023.

United Nations Economic and Social Council. (1948). *Resolution of 11 February 1948: United Nations Scientific Conference on the Conservation and Utilisation of Resources*. E/RES/109(VI). Available at: https://documents-dds-ny.un.org/doc/RESOLUTION/GEN/NR0/072/44/img/NR007244.pdf?OpenElement. Accessed 19 April 2023.

United Nations Economic and Social Council. (1969). *Economic and Social Council Official Records: Forty-Seventh Session, 14 July–8 August 1969*. Available at: https://documents-dds-ny.un.org/doc/UNDOC/GEN/NL1/490/66/pdf/NL149066.pdf?OpenElement. Accessed 20 April 2023.

United Nations Economic Commission for Europe. (1991). *Conclusions of the Conference 'Environment for Europe', Dobříš Castle, Czech and Slovak Republic, June 21–23, 1991*. Available at: https://unece.org/fileadmin/DAM/env/efe/history%20of%20EfE/Dobříš.E.pdf. Accessed 30 April 2023.

United Nations Economic Commission for Europe. (1993). *Declaration by the ministers of the environment of the region of the United Nations Economic Commission for Europe (UN/ECE) and the member of the Commission of the European Communities responsible for the environment*. Available at: https://unece.org/fileadmin/DAM/env/efe/history%20of%20EfE/Luzern.E.pdf. Accessed 30 April 2023.

United Nations Economic Commission for Europe. (1995a). *Declaration by the ministers of environment of the region of the United Nations Economic Commission for Europe*. Available at: https://unece.org/fileadmin/DAM/env/efe/history%20of%20EfE/Sofia.E.pdf. Accessed 31 March 2023.

United Nations Economic Commission for Europe. (1995b). *Draft Guidelines on Access to Environmental Information and Public Participation in Environmental Decision-making*. ECE/CEP/24. Available at: https://unece.org/DAM/env/documents/1995/cep/ece.cep.24e.pdf. Accessed 1 April 2023.

United Nations Economic Commission for Europe. (1996a). *Draft elements for the convention on access to environmental information and public participation in environmental decision-making*. CEP/AC.3/R.1. Available at: https://unece.org/DAM/env/documents/1996/cep/ac.3/cep.ac3.r1.e.pdf. Accessed 1 April 2023.

United Nations Economic Commission for Europe. (1996b). *Economic Commission for Europe Committee on Environmental Policy: Report on the Special Session, 17 January 1996*. ECE/CEP/18. Available at: https://unece.org/DAM/env/documents/1996/cep/ece.cep.18.e.pdf. Accessed 1 April 2023.

United Nations Economic Commission for Europe. (1996c). *Report of the first session: Working group for the preparation of a draft convention on access to environmental information and public participation in environmental decision-making*. CEP/AC.3/2. Available at: https://unece.org/DAM/env/documents/1996/cep/ac.3/cep.ac.3.2.e.pdf. Accessed 1 April 2023.

United Nations Economic Commission for Europe. (1996d). *Report of the second session: Working group for the preparation of a draft convention on access to environmental information and public participation in environmental decision-making*. CEP/AC.3/4. Available at: https://unece.org/DAM/env/documents/1996/cep/ac.3/cep.ac.3.4.e.pdf. Accessed 1 April 2023.

United Nations Economic Commission for Europe. (1997a). *Report of the eighth session: Working group for the preparation of a draft convention on access to environmental information and public participation in environmental decision-making*. CEP/AC.3/16. Available at: https://unece.org/DAM/env/documents/1997/cep/ac.3/cep.ac.3.16.e.pdf. Accessed 2 April 2023.

United Nations Economic Commission for Europe. (1997b). *Report of the fifth session: Working group for the preparation of a draft convention on access to environmental information and public participation in environmental decision-making*. CEP/AC.3/10. Available at: https://unece.org/DAM/env/documents/1997/cep/ac.3/cep.ac.3.10.e.pdf. Accessed 2 April 2023.

United Nations Economic Commission for Europe. (1997c). *Report of the fourth session: Working group for the preparation of a draft convention on access to environmental information and public participation in environmental decision-making*. CEP/AC.3/8. Available at: https://unece.org/DAM/env/documents/1997/cep/ac.3/cep.ac.3.8.e.pdf. Accessed 2 April 2023.

United Nations Economic Commission for Europe. (1997d). *Report of the seventh session: Working group for the preparation of a draft convention on access to environmental information and public participation in environmental decision-making*. CEP/AC.3/14. Available at: https://unece.org/DAM/env/documents/1997/cep/ac.3/cep.ac.3.14.e.pdf. Accessed 2 April 2023.

United Nations Economic Commission for Europe. (1997e). *Report of the sixth session: Working group for the preparation of a draft convention on access to environmental information and public participation in environmental decision-making*. CEP/AC.3/12. Available at: https://unece.org/DAM/env/documents/1997/cep/ac.3/cep.ac.3.12.e.pdf. Accessed 2 April 2023.

United Nations Economic Commission for Europe. (1997f). *Report of the third session: Working group for the preparation of a draft convention on access to environmental information and public participation in environmental decision-making*. CEP/AC.3/6. Available at: https://unece.org/DAM/env/documents/1997/cep/ac.3/cep.ac.3.6.e.pdf. Accessed 2 April 2023.

United Nations Economic Commission for Europe. (1998a). *Declaration by the ministers of environment of the region of the United Nations Economic Commission for Europe (UN/ECE)*. ECE/CEP/56. Available at: https://unece.org/fileadmin/DAM/env/efe/history%20of%20EfE/Aarhus.E.pdf. Accessed 4 April 2023.

United Nations Economic Commission for Europe. (1998b). *Report of the ninth session: Working group for the preparation of a draft convention on access to environmental information and public participation in environmental decision-making*. CEP/AC.3/18. Available at: https://unece.org/DAM/env/documents/1998/cep/ac.3/cep.ac.3.18.e.pdf. Accessed 2 April 2023.

United Nations Economic Commission for Europe. (1998c). *Report of the tenth session: Working group for the preparation of a draft convention on access to environmental information and public participation in environmental decision-making*. CEP/AC.3/20. Available at: https://unece.org/DAM/env/documents/1998/cep/ac.3/cep.ac.3.20.e.pdf. Accessed 2 April 2023.

United Nations Economic Commission for Europe (1998d) Report on the fourth ministerial conference 'Environment for Europe', Aarhus, Denmark, 23–25 June 1998. ECE/CEP/41. Available at: https://unece.org/fileadmin/DAM/env/documents/1998/ece/ece.cep.41.e.pdf. Accessed 4 April 2023.

United Nations Economic Commission for Europe. (2000). *The Aarhus Convention: An implementation guide*. United Nations Economic Commission for Europe.

United Nations Economic Commission for Europe. (2005). *Report of the second meeting of the parties*. ECE/MP.PP/2005/2. Available at: https://unece.org/fileadmin/DAM/env/documents/2005/pp/ece/ece.mp.pp.2005.2.e.pdf. Accessed 19 May 2023.

United Nations Economic Commission for Europe. (2022). *Report of the seventh session of the Meeting of the Parties*. ECE/MP.PP/2021/2. Available at: https://unece.org/sites/default/files/2022-07/ECE_MP.PP_2021_2_E.pdf. Accessed 17 May 2023.

United Nations Economic Commission for Europe. (2023a). *Environment for Europe*. Available at: https://unece.org/environment-policy/environment-europe. Accessed 28 March 2023.

United Nations Economic Commission for Europe. (2023b). *Structure*. Available at: https://unece.org/structure. Accessed 28 March 2023.

United Nations Economic Commission for Europe. (2023c). *UNECE archives: Reports on the negotiations of the Aarhus Convention*. Available at: https://unece.org/environment-policy/public-participation/aarhus-convention/reports-negotiations-convention. Accessed 6 April 2023.

United Nations Educational, Scientific and Cultural Organisation. (1950). *International technical conference on the protection of nature: Lake success, 22–29, VIII, 1949*. Available at: https://portals.iucn.org/library/sites/library/files/documents/1950-001-pg_i-xi.pdf. Accessed 19 April 2023.

United Nations Educational, Scientific and Cultural Organisation. (1993). *The biosphere conference, 25 years later*. UNESCO. Available at: https://unesdoc.unesco.org/ark:/48223/pf0000147152. Accessed 18 April 2023.
United Nations Educational, Scientific and Cultural Organisation. (2021). *50th Anniversary of UNESCO's Man and the Biosphere Programme*. Available at: https://en.unesco.org/mab/50years. Accessed 20 April 2023.
United Nations Educational, Scientific and Cultural Organisation. (2022). *Mapping the evolving interface of access to information and privacy: Issue brief*. Available at: https://unesdoc.unesco.org/ark:/48223/pf0000384099. Accessed 16 April 2023.
United Nations Educational, Scientific and Cultural Organisation. (2023). *Video and sound collections: Scientific basis for rational use and conservation of the biosphere*. Available at: https://www.unesco.org/archives/multimedia/search/scientific+basis+for+rational+use. Accessed 20 April 2023.
United Nations Environment Programme. (1987). *Goals and principles of environmental impact assessment*. Available at: https://digitallibrary.un.org/record/42521?ln=en. Accessed 21 April 2023.
United Nations Environment Programme. (2023). *The Montreal protocol*. Available at: https://www.unep.org/ozonaction/who-we-are/about-montreal-protocol. Accessed 21 April 2023.
United Nations General Assembly. (1946). *Resolution 59(I). Calling of an international conference on freedom of information*. A/RES/59(I). Available at: https://documents-dds-ny.un.org/doc/RESOLUTION/GEN/NR0/033/10/PDF/NR003310.pdf?OpenElement. Accessed 15 April 2023.
United Nations General Assembly. (1962). *Resolution 1840(XVII). Draft Convention on Freedom of Information; Draft Declaration on Freedom of Information*. A/RES/1840(XVII). Available at: https://digitallibrary.un.org/record/204348?ln=en. Accessed 16 April 2023.
United Nations General Assembly. (1968). *Resolution 2398(XXIII). Problems of the human environment*. Available at: https://documents-dds-ny.un.org/doc/RESOLUTION/GEN/NR0/243/58/img/NR024358.pdf?OpenElement. Accessed 20 April 2023.
United Nations General Assembly. (1969). *Resolution 2542(XXIV). Declaration on Social Progress and Development*. A/RES/2542(XXIV). Available at: https://digitallibrary.un.org/record/202280?ln=en. Accessed 16 April 2023.
United Nations General Assembly. (1974). *Freedom of information: Note by the secretary-general*. A/9657. Available at: https://digitallibrary.un.org/record/855613?ln=en. Accessed 16 April 2023.
United Nations General Assembly. (1982). *Resolution 37/7. World charter for nature*. A/RES/37/7. Available at: https://digitallibrary.un.org/record/39295?ln=en. Accessed 21 April 2023.

United Nations General Assembly. (1989). *Resolution 44/228. UN Conference on Environment and Development*. A/RES/44/228. Available at: http://www.un-documents.net/a44r226.htm. Accessed 28 March 2023.
United Nations General Assembly. (1990). *Resolution 45/94. Need to ensure a healthy environment for the well-being of individuals*. A/RES/45/94. Available at: https://digitallibrary.un.org/record/105298?ln=en. Accessed 25 April 2023.
United Nations General Assembly. (2011). *Report of the International Law Commission, Sixty-third Session*. A/66/10. Available at: https://documents-dds-ny.un.org/doc/UNDOC/GEN/N11/527/74/pdf/N1152774.pdf?OpenElement. Accessed 6 April 2023.
United Nations High Commissioner for Human Rights. (2011). *Report of the United Nations High Commissioner for Human Rights: Analytical study on the relationship between human rights and the environment*. A/HRC/19/34. Available at: https://www.ohchr.org/sites/default/files/Documents/HRBodies/HRCouncil/RegularSession/Session19/A-HRC-19-34_en.pdf. Accessed 25 April 2023.
United Nations Treaty Collection. (2023). *Multilateral treaties deposited with the Secretary-General*. Available at: https://treaties.un.org/Pages/Index.aspx?clang=_en. Accessed 15 April 2023.
Universal Declaration of Human Rights. (1948). A/RES/217(III). Available at: https://www.un.org/en/development/desa/population/migration/generalassembly/docs/globalcompact/A_RES_217(III).pdf. Accessed 15 April 2023.
Vienna Convention on the Law of Treaties. (1969). *United Nations Treaty Series, 1155, 331*. Available at: https://legal.un.org/ilc/texts/instruments/english/conventions/1_1_1969.pdf. Accessed 6 April 2023.
Waltz, K. (1979). *Theory of international politics*. McGraw-Hill.
Wates, J. (2005). The Aarhus Convention: A driving force for environmental democracy. *Journal of European Environmental and Planning Law, 1*, 2–11.
Wates, J. (2011). The future of the Aarhus Convention: Perspectives arising from the third session of the Meeting of the Parties. In M. Pallemaerts (Ed.), *The Aarhus Convention at ten: Interactions and tensions between conventional international law and EU environmental law* (pp. 383–415). Europa Law.
Weaver, D. (2015). *The Aarhus Convention: towards a cosmopolitan international environmental politics* (Unpublished PhD thesis). Keele University. Available at: https://eprints.keele.ac.uk/id/eprint/2310/1/Weaverphd2015.pdf. Accessed 12 April 2023.

Weaver, D. (2020). Spatiality and world politics. In N. Sandal (Ed.), *Oxford research encyclopedia of international studies* (pp. 1–37). Available at: https://oxfordre.com/internationalstudies/display/10.1093/acrefore/9780190846626.001.0001/acrefore-9780190846626-e-562?print=pdf. Accessed 11 April 2023.

Weinert, M. S. (2007). Bridging the human rights-sovereignty divide: Theoretical foundations of a democratic sovereignty. *Human Rights Review, 2007*, 5–32.

Weinthal, E., & Watters, K. (2010). Transnational environmental activism in Central Asia: The coupling of domestic law and international conventions. *Environmental Politics, 19*(5), 782–807.

Werner, W. (2016). State consent as foundational myth. In C. Brölmann & Y. Radi (Eds.), *Research handbook on the theory and practice of international lawmaking* (pp. 13–31). Edward Elgar.

Whitton, J. B. (1949). The United Nations conference on freedom of information and the movement against international propaganda. *American Journal of International Law, 43*(1), 73–87.

World Commission on Environment and Development. (1987). *Our common future*. Oxford University Press.

World Health Organisation. (1989). *European Charter on Environment and Health [Online]*. Available at: http://www.euro.who.int/__data/assets/pdf_file/0019/114085/ICP_RUD_113.pdf. Accessed 30 October 2012.

Yanitsky, O. N. (2012). From nature protection to politics: The Russian environmental movement, 1960–2010. *Environmental Politics, 21*(6), 922–940.

Zaharchenko, T. (2009). *On the way to transparency: A comparative study on post-soviet states and the Aarhus Convention [Online]*. Available at: https://www.wilsoncenter.org/sites/default/files/media/documents/publication/op303_on_way_to_transparency_zaharchenk_2009.pdf. Accessed 12 April 2023.

Zaharchenko, T. R., & Goldenman, G. (2004). Accountability in governance: The challenge of implementing the Aarhus Convention in Eastern Europe and Central Asia. *International Environmental Agreements, 4*, 229–251.

Ziegler, C. E. (1987). *Environmental policy in the USSR*. University of Massachusetts Press.

Zillman, J. (2009). *A history of climate activities*. Available at: https://public.wmo.int/en/bulletin/history-climate-activities. Accessed 26 April 2023.

Zubok, V. (2022). *Collapse: The fall of the Soviet Union*. Yale University Press.

CHAPTER 4

Germination: The Aarhus Convention's Procedural Trinity

Introduction[I]

This chapter explores Aarhus' (Aarhus Convention, 1998) trinity of procedural rights to information access, public participation and access to justice in environmental decision-making. Each is a Greek pillar: it exists in a mutually reinforcing relationship with the others. The first pillar, information access, is provided for in Articles 4 and 5. It is *primus inter pares*: the first prerequisite of "good governance" and a precondition for effective public participation and pursuit of justice. The second pillar, public participation, is addressed in Articles 6, 7 and 8. It is an outcome of the effective use of accurate information, and an enabler of the pursuit of justice. In turn, the third pillar, access to justice, is provided for in Article 9. It is both an outcome and guardian of the two preceding pillars (Weaver, 2015, 2018).

With this in mind, affinities exist between Aarhus' procedural trinity and a valuable concept in the environmental communication literature, the trinity of voice (Cox, 2006: 142–145; Senecah, 2004). The latter incorporates access, standing and influence. Access, firstly, entails here more than information access. It includes concerns for collaboration, facilitation of participatory opportunity and human involvement, transparency and education (Senecah, 2004: 23–24). It is thus premised on what I would regard as a cosmopolitan impulse of human empowerment. Access, in the trinity of voice, requires a positive disposition towards

D. Weaver, *The Aarhus Convention*, Environmental Politics and Theory,
https://doi.org/10.1007/978-3-031-43536-2_4

co-creation, and sufficient empowerment to ensure that humans can effectively contribute to decision-making procedures. This understanding of access overlaps with Aarhus' first and second pillars.

Standing in the trinity of voice, meanwhile, does not denote the locus standi of litigation, but more broadly embraces "dialogue and deliberation; active listening...early and ongoing voice" plus "genuine empathy for the concerns of other perspectives" (Senecah, 2004: 24). This notion of standing is well aligned with Aarhus' second pillar. Finally, influence entails the respectful consideration of human contributions in concert with those of state and private sector actors (Senecah, 2004: 25). Influence denotes "the element felt by many to be most often missing" (Cox, 2006: 142) in environmental decision-making. Influence surfaces in "meaningful decision space, transparent process that considers all alternatives, opportunities to meaningfully scope alternatives, opportunities to inform the decision criteria, and thoughtful response to stakeholder concerns and ideas" (Senecah, 2004: 25). This again is well aligned with Aarhus' second pillar, not least given the latter's due account provisions (see below).

As this chapter will demonstrate, Aarhus' procedural rights help to operationalise the trinity of voice. They help "amplify" public environmental voice and set the circumstances in which humans and their associations can more ably articulate themselves and their environmental concerns. Aarhus makes modest progress towards the institutionalisation of state-public co-creation. But I stress modesty here. Aarhus will be seen, in this chapter, to mildly enhance human abilities to be "heard" in environmental decision-making. The focus is on attuning states to the expression of human and environmental interests, in such a way that existing power arrangements are reinforced rather than usurped.

What follows thus coheres with Richardson and Razzaque's (2006: 173; emphasis added) finding that entrenchments of environmental public participation "hardly seem to threaten existing political institutions, since they operate *within* those institutions, and leave power and authority *mostly unfettered*". Their apt intervention reinforces this book's identification of solidarisation, underway in the Convention, which posits that any radical solidarist change necessarily needs framing *inside* the existing power arrangements of international society, in order to remain palatable for consenting states (see Ahrens, 2017, 2018, 2019; Ahrens & Diez, 2015).

Of course, please do not read this as playing down, too firmly, the ethical significance of Aarhus' procedural trinity. What follows should constitute a substantiation of Pallemaerts' (2011: 4) view that Aarhus helps cast environmental policy as "a testing ground for efforts to transcend traditional models of representative democracy". Its procedural rights embolden those seeking environmental protection in the context of democratic reform. Its provisions presuppose an "*engaged, critically aware public*" that plays a dual role of "*player* and...*partner* in the formulation and implementation of environmental policies" (Pallemaerts 2011: 4; emphasis added). Pallemaerts' "player-partner" understanding speaks to co-creation: rich, dialogic partnership between state and human, with the latter benefiting from some greater or lesser parity of actorhood in influencing decision-making procedures and outcomes. Co-creation would climb highly on Arnstein's (1969) somewhat clichéd ladder of public participation. As players, the public would contribute palpably to decision *outcomes*, contributing their knowledge, expertise, concern and ethical outlook in order to enrich an outcome that is more reflective of human diversity, more holistically informed, and—one hopes—more environmentally benign. As partners, the public would be more invested in enhancing the democratic health and procedural justice of decision-making *processes*.

To perform this dual player-partner role, the public needs accurate information, relevant to the decision-making procedures in question. Information must be accessible, comprehensible and in a language spoken and understood by the participants (Weaver, 2018). Information should be withheld only in the most extreme circumstances. The public needs the capacity to participate without fear of chastisement or retribution, safe in the knowledge that their contributions—of lay, indigenous knowledge and also of Western scientific "expertise" (see Weaver, 2023; Reed, 2008: 2425)—will be sincerely heard and accounted for. Finally, the public will only yield influence to the extent that it can pursue redress when its rights are violated. There is no utility in accountability-holding if it is not justiciable.

This is the logic that I perceive to animate Aarhus' procedural trinity. The logic resonates with Gilmore's (2023) solidarist work on responsible statehood, concerned for a co-production that "opens up" state governance practices to nonstate actors in more equitable dialogue. Such "opening up" requires states to develop "practices that can engage more directly with publics both within and transecting the borders of their state" (Gilmore, 2023: 97). The effectiveness of such "opening up" lies in the impartiality with which it occurs. Provision of Aarhus' rights—no, provision of *all* human rights—must not discern between recipients. It would be a cruel injustice to create hierarchies of rights-holders. Hierarchies would perpetuate power asymmetries, paving another hurtful path to discrimination and dehumanisation. Such hierarchies would enable the power*ful* to again exploit their dominant position vis-à-vis the power*less*. Seeking to avoid such injustice, a strict moral cosmopolitanism is needed, starting from the position that one human life, or at least one instance of human agency, is worth one human life or instance thereof. It is crucial that, in decision-making procedures (particularly environmental procedures likely to harm human and environmental health), *all* humans receive rights equally. This helps to avoid corruption, prevents favourable treatment of the power*ful* and privileged and helps to safeguard against the silencing of the power*less* and underprivileged. Silencing, after all, constitutes violence per se, not least in environmental contexts (Brisman, 2013).

This logic of human equality is simple yet profound. And it can be extremely disquieting, given humans' frightening capacity to harm human and nonhuman nature, and likely lay responses that those culpable for the most egregious harms somehow forfeit their rights. But to me at least, a pure moral cosmopolitanism accepts, warts and all, humans' total equality "at dawn and at night", irrespective of their polarisations and vices. Fundamental human rights are "fundamental" and "human" with good reason. They apply to each human equally, without distinction or discrimination. This logic is strictly morally cosmopolitan, and I do not for one moment suggest that it is always borne in reality. But nonetheless, this is the vantage point from which I study international ethics, and which animates my understanding of a richer, "humankind" solidarism (see Chapter 2).

With this in mind, the "opening up" of state governance practices and the enshrinement of co-creation is conducive to the ancient all-affected principle, whose logic is that what touches all should be heard and approved by all (Linklater, 2002: 146; see Ball, 2006; Eckersley, 2004: 112; Warren, 2017). I argue that Aarhus operationalises the all-affected principle, not least given its provision that the public (Aarhus Convention, 1998, Article 3.9; emphasis added) "shall have access to information, have the possibility to participate in decision-making and have access to justice in environmental matters ***without discrimination as to citizenship, nationality or domicile*** and, in the case of a legal person, without discrimination as to where it has its registered seat or an effective centre of its activities". This provision helps to ease the national/foreigner distinction, levelling the "pitch" on which humans receive their rights, and making progress towards the moral cosmopolitanism above.

What also needs bearing in mind, before we assess each pillar, is Aarhus' provision of rights to the *public* and *public concerned*. Article 2.4 (Aarhus Convention, 1998) defines the public as "one or more natural or legal persons, and, in accordance with national legislation or practice, their associations, organisations or groups". This implies the "any person" principle; "associations, organisations or groups" will tend to already have legal personality and meet this definition (Ebbesson et al., 2014: 55).

Furthermore, the definition of "public" is not bound by restrictions. As such, where Aarhus assigns rights to the public without caveat, the public can exercise them irrespective of whether they are affected by, or have any interest in, the decision-making in question (Ebbesson et al., 2014: 55). This broad notion of "public" is applied with respect to Aarhus' information provisions in Articles 4 and 5. For this reason, "any person" (Aarhus Convention, 1998, Article 9.1; see this chapter, below) receives access to justice rights where Parties' compliance with Aarhus' first pillar information rights is called into question. Such broad use of "public" is also (see below) used in Articles 6.7 (on public submission of comments, information, analyses or opinions relevant to proposed activities), 6.9 (on promptly informing the public of decisions) and 8 (on public participation during preparation of executive regulations or generally applicable legally binding normative instruments) (Aarhus Convention, 1998).

Additionally, and keeping this broad "any person" approach to the public in mind, I should also signpost the reader to Article 9.3 (Aarhus Convention, 1998), which pushes towards *actio popularis* in assigning access to justice rights to "members of the public" in instances of national environmental law being allegedly contravened. This is particularly empowering, and presents the possibility for the human to benefit from environmental legal actorhood. And recalling the earlier national/foreigner nullification, it matters not whether the public applying these rights are Parties' citizens. This helps actualise the above moral cosmopolitanism that I deem to ethically drive solidarism.

If Article 2.4 defines the *public*, Article 2.5 (Aarhus Convention, 1998) defines the *public concerned* as "the public affected or likely to be affected by, or having an interest in, the environmental decision-making; for the purposes of this definition, non-governmental organisations promoting environmental protection and meeting any requirements under national law shall be deemed to have an interest". The concept of the public concerned is used to define the legal persons entitled to exercise Article 6 participation rights (on specific environment-related activities) and hence Article 9.2 rights granting access to justice in cases of Parties' non-compliance with Article 6 duties and "other relevant" Convention provisions (Aarhus Convention, 1998; see this chapter, below). It has been noted that this idea of the public concerned employs "the concept of "being affected" which is well known in some jurisdictions and has been used since 1991 by the Espoo Convention for the purpose of defining the public which should be allowed to participate in transboundary EIA" (Ebbesson et al., 2014: 57). Aarhus contributes, here, to an operationalisation of the all-affected principle, as the public concerned only need *factual* interests in given decision-making procedures, rather than specific *legal* interests (Ebbesson et al., 2014: 57).

Finally, the provision for NGOs to meet "any requirements under national law" should not be construed as giving carte blanche to states to wholesale curtail NGOs' locus standi. Whilst Parties *can* set requirements for NGOs' standing in domestic law, requirements should neither be politically charged nor too onerous; it is in Aarhus' spirit that Parties encourage civil society and its contribution to governance (Ebbesson et al., 2014: 58; see Aarhus Convention, 1998).

What I hope to have conveyed, in this opening coverage, is that Aarhus' procedural trinity is driven by a cosmopolitan impulse of human empowerment and co-creation, with humans being invested not only in the ethics of environmental protection but also in domestic and international societies' democratic health and civic responsibility. Through English School lenses, such co-creation surely values the enhanced role of human world society actors "as legitimate stakeholders and even partners in the crafting of global environmental solutions" (Falkner, 2021: 273). Keeping those lenses in, let us now explore the trinity and further excavate those English School implications.

The First Pillar: Access to Information

Access to information, the first pillar in Aarhus' trinity, is "powerful and detailed" (Blundell, 2015: 99). It is *primus inter pares*, given that public participation and the pursuit of justice are infeasible without accurate information, obtained without hindrance and provided in the best of faith by relevant actors. The first pillar safeguards commitments to transparency, enabling accountability-holding by environmentally and democratically engaged humans and their associations. The logic is that to be "blind" to the detail and implications of a decision is to be disempowered. Knowledge is power. Information thus needs to be as accessible and comprehensible as possible. If it is masked in technical jargon, it will remain a luxury used only by participatory elites whose contributions may not align with potentially affected persons; this would constitute injustice (Weaver, 2018). Moreover, citizen empowerment will fail unless publics *en masse* receive the knowledge with which to participate effectively. Sincere commitment, by states, to safeguarding equity, diversity and inclusion (EDI) in environmental decision-making procedures is necessary.

It is, furthermore, important that Parties discharge their duties to provide "information about information" (Aarhus Convention, 1998, Article 5.2a) as well as to support public actors in their search for information (Aarhus Convention, 1998, Article 5.2bii). Just as higher education (HE) witnessed ethically proper and socially just widening participation efforts, enabling the achievement of persons who would erstwhile not have contemplated HE, *environmental education* also requires such efforts. First generation environmental decision-making participants need the same capacity building as first generation HE students. The ethical impulse is the same: it is one of human empowerment.

Having cleared preliminary ground, let us turn to the first pillar's particulars. I should stress that in lieu of a substantive definition of environment, Aarhus (Aarhus Convention, 1998, Article 2.3) uses a broad, non-exhaustive definition of environmental information that can take "written, visual, aural, electronic or any other material form". Environmental information can relate to the "state of elements of the environment" including air, atmosphere, water, soil, land, natural sites, biological diversity, GMOs and the interplay amongst them (Aarhus Convention, 1998, Article 2.3a). It can also concern such factors as substances, energy, noise and radiation, alongside such activities and measures as administrative measures, environmental agreements, policies, legislation, plans and programmes, which either affect or are likely to affect elements of the environment specified in Article 2.3 (Aarhus Convention, 1998). It can yet further include the state of human health and safety, conditions of human life, cultural sites and built structures, to the extent that they are (or may be) affected by the state of the environment, or by the above phenomena (Aarhus Convention, 1998, Article 2.3c).

Holding this broad definition in mind, Aarhus' first pillar has *passive* (Aarhus Convention, 1998, Article 4) and *active* (Aarhus Convention, 1998, Article 5) provisions. Article 4 provisions are passive because, under them, it is incumbent on the public to make an information request, and for the relevant public authority to respond. As stated above, all persons have the right to access environmental information. When a request is made, it is to be provided "within the framework of national legislation" (Aarhus Convention, 1998, Article 4.1). Whilst public authorities can limit information disclosure given the presence of exemptions (see below), states must ensure that their domestic law governs their ability to respond to information requests in a manner compliant with the Convention (Ebbesson et al., 2014: 79). The pillar thus does not give states carte blanche to arbitrarily refuse information requests. It rather inclines Aarhus' state Parties from *need to know* to *right to know* information governance modalities (see Cramer, 2009; Florini, 2007; Gavouneli, 2000; Hallo, 2011; Harman-Stokes, 1995; Krämer, 2012; Mason, 2010; Sand, 2003). This is evident in the provision that no interest needs to be stated by the person requesting the information (Aarhus Convention, 1998, Article 4.1a). Such a provision is firmly in line with the "any person" principle (Wates, 2005: 3), and requests cannot be rejected because the applicant does not have, or does not state, an interest (Ebbesson et al., 2014: 80).

Moreover, time is of the essence in the passive element. Public authorities must provide requested information ordinarily within a month, unless its complexity and volume prevent this, in which case two months are permitted (Aarhus Convention, 1998, Article 4.2). If the public authority does not hold the information, it must "as promptly as possible" either advise the applicant of the authority that *does* hold it, or "transfer the request to that authority and inform the applicant accordingly" (Aarhus Convention, 1998, Article 4.5). This latter direct transfer provision adopts a spirit of partnership, firmly oriented around human empowerment. It begins from a position of deliberate help as opposed to hindrance, and should be understood in the earlier context of democratisation (see this book, Chapter 3).

However, such a cosmopolitan impulse is counterbalanced by a suite of exemptions that enable Parties to deny information requests. Requests can be denied under Article 4.3 (Aarhus Convention, 1998) if the public authority does not hold the information; if the request is "manifestly unreasonable or formulated in too general a manner"; if the request concerns material in the course of completion; or if the request concerns authorities' internal communications subject to exemptions provided in law or custom, "taking into account the public interest served by disclosure". "Material in the course of completion" denotes drafts being actively worked on; "internal communications" protect, in some countries, personal opinions of government staff (Ebbesson et al., 2014: 85). The crux is that if consideration is given to exempting drafts or materials concerning internal communications, the public interest in disclosure still needs accounting for. The public interest test (Aarhus Convention, 1998, Article 4.3c) seeks to "prevent abuse of the exemptions by over-secretive public authorities" (Wates, 2005: 4; see Blundell, 2015: 107–108). It constitutes an added safeguard against Parties that might not be wholly committed to environmental accountability. The point is that states do not have total discretion to arbitrarily deny information requests. If the public interest is met, then the information still needs to be disclosed. The "right to know" takes precedence over the "need to know".

Further exemptions in Article 4.4 (Aarhus Convention, 1998) are distinct as they carry with them stated interests, and it is incumbent on the relevant authority to determine that those interests will be jeopardised, in order for the exemption to be valid (see Ebbesson et al., 2014: 86). Requests can be denied under this provision (Aarhus Convention, 1998, Article 4.4) if they would adversely affect the confidentiality of public authorities' proceedings; international relations, defence or security; the course of justice; commercial and industrial confidentiality; intellectual property rights; the confidentiality of personal data where disclosure consent was not gained; the interests of third parties that voluntarily supplied information without disclosure consent; and the environment *itself*, for instance where breeding site locations might be compromised by disclosure.

The goal, here, is to protect confidentiality in the most sensitive of circumstances. The provision on the confidentiality of public authorities' proceedings will already be amply accommodated by advanced democracies' domestic laws. From a UK vista, the Official Secrets Acts are relevant and, as I write, are being assessed for their continued relevance with a view to reform (Home Office, 2023). UK Government Security Classifications (Cabinet Office, 2018) and the Data Protection Act (2018) are also relevant. Advanced democracies will already have robust measures with which to safeguard the integrity of state information. There are meticulous frameworks in place to govern the conditions in which state information is, and is not, disclosed to the public.

Meanwhile, and turning to international relations, defence and security, it goes without saying that states will necessarily withhold information of such sensitivity that disclosure would compromise their national interests, particularly those relating to their military and security interests. Aarhus' Parties would never have consented to an information access pillar that undermined their own *raison d'état*. Similarly, the course of justice, personal and commercial confidentiality, intellectual property, the voluntary conveyance of information by individuals to governments, and highly sensitive environmental data each merit concern for ensuring the safety not only of the *information*, but also of its *subjects* and *sources*. Good reason exists to justify why some information must be withheld, for the security not only of states and populations, but also for the preservation of human and environmental rights.

Nevertheless, if requested information can be extracted from otherwise exempted content, it *should* be (Aarhus Convention, 1998, Article 4.6). This reaffirms the pillar's turn towards "right to know" governance modalities. Refusals should be justified, and information on review procedures (see Aarhus Convention, 1998, Article 9; see also this chapter, below) provided (Aarhus Convention, 1998, Article 4.7). Refusals are to be made "as soon as possible and at the latest within one month", unless the information's complexity justifies two months (Aarhus Convention, 1998, Article 4.7). Whilst disclosure may be chargeable, charges need to be reasonable (Aarhus Convention, 1998, Article 4.8). This is a further modest contribution towards human empowerment. It seeks to offer the proverbial tools, and set the favourable conditions, for effective public participation in environmental decision-making procedures.

Having addressed the passive element, let us now turn to the active element of the first pillar. Three themes are elicited from it: transparency, centralisation of information and consumer rights. The latter might appear to be a curveball, but should be understood in the above contexts of market-political liberalisation and democratisation (see this book, Chapter 3). Noting that the above exemptions are reaffirmed in the active element (Aarhus Convention, 1998, Article 5.10), let me address each theme in turn.

With regard to transparency, firstly, the most important provision is the requirement for Parties to submit national "state of the environment" reports "at regular intervals not exceeding three or four years" (Aarhus Convention, 1998, Article 5.4). It is noteworthy that such reporting was regarded as an effective enabler of environmental accountability by scholars (Eckersley, 2000: 129; Saward, 2006: 188). Concern for transparency is also evident in the provision that, during environmental emergencies, active dissemination of environmental information must occur "immediately and without delay to members of the public who may be affected" (Aarhus Convention, 1998, Article 5.1c). The focus, there, is on saving life and preventing immediate harm. Such emergency dissemination will occur when human or environmental health faces imminent threat, of such severity that the exemptions will likely be overridden (Blundell, 2015: 115, 119). This is a clear indicator of the all-affected principle being operationalised, because it breathes practical legal life into the position that environmental harm that "touches all" should be "heard by all": it renders state actors responsible for disseminating the information necessary for humans to make informed choices about situations likely to harm them and their environments.

When information is actively disseminated to the public, it should be done so accessibly and transparently (Aarhus Convention, 1998, Article 5.2). It will likely relate to materials concerning (inter)national law and policy (Aarhus Convention, 1998, Article 5.5), plus environmental policymaking, Parties' dealings with the public under the Convention, and government provision of environmental functions and services (Aarhus Convention, 1998, Article 5.7). These active requirements seek to ensure that transparency is safeguarded and trust is built between state and public. The greater the trust, the greater the chances that the public will feel inclined and able to contribute to decision-making procedures, playing that earlier dual role of player-partner (Abels, 2007; Böhmelt et al., 2014; Klinke, 2012; Reed, 2008; Richardson & Razzaque, 2006: 170; Spyke, 1999; Steele, 2001).

The second theme, in the active element, concerns centralisation of information. This is evident in the requirement for public authorities to possess and update environmental information relevant to their functions (Aarhus Convention, 1998, Article 5.1a). It is also evident in the requirement to create mandatory systems to ensure adequate flow of information to authorities about activities that may significantly impact the environment (Aarhus Convention, 1998, Article 5.1b). Electronic databases for environmental information are mandated (Aarhus Convention, 1998, Article 5.3), as are electronic pollution registers (Aarhus Convention, 1998, Article 5.9). The latter materialised when Parties agreed the Protocol on Pollutant Release and Transfer Registers (the PRTR Protocol) (Protocol on Pollutant Release and Transfer Registers to the Convention on Access to Information, Public Participation in Decision-Making and Access to Justice in Environmental Matters) in 2003. The PRTR Protocol entered into force in 2009. This logic of centralisation is again premised upon a concern for human empowerment: centralised information repositories should make it easier for states and their public authorities to (a) monitor and keep track of their environment and hence (b) facilitate effective human engagement in environmental decision-making practices.

The third theme, consumer rights, might initially appear to be a curveball. But it should be understood in the earlier context of liberalisation and democratisation (see this book, Chapter 3). Parties are required to encourage "operators" (private enterprises and other actors whose activities have significant environmental impact) "to inform the public regularly of the environmental impact of their activities and products, where appropriate within the framework of voluntary eco-labelling or eco-auditing schemes or by other means" (Aarhus Convention, 1998, Article 5.6). Parties are also required to "develop mechanisms with a view to ensuring that sufficient product information is made available to the public in a manner which enables consumers to make informed environmental choices" (Aarhus Convention, 1998, Article 5.8). This constitutes another take on human empowerment, seeking to endow humans-as-consumers with enhanced capacities to make informed environmental choices in an arena of free market choice saturation. In the aftermath of environmentally destructive "command-and-control" state socialism, alongside the risks of "runaway" (and arguably equally injurious) post-socialist privatisation (Yanitsky, 2012), it makes sense why Aarhus would entail the attempted greening of corporate responsibility and consumer rights.

Let me now distil some thoughts on the first pillar. It embodies a manoeuvre from "need to know" to "right to know" information governance modalities, garnering a spirit of state-public co-creation. The underlying cosmopolitan impulse is one of human empowerment, visible in the fact that persons making information requests need not disclose an interest in the information sought. It is evident in the pillar's affinity with, and operationalisation of, the "all-affected principle", requiring that all potentially affected persons receive the knowledge needed to make informed choices, and to influence developments that might harm them. The provisions for environmental emergencies and mandatory reporting are but two instances, moreover, of the "right to know" being invoked. State obligations relating to timeliness of disclosure, reasonable charges, public assistance and direct transfer of information requests "nudge" state actors in the direction of co-creation.

As such, rather than *emancipating* humans, the pillar *empowers* them by providing the proverbial tools needed to yield influence in environmental decision-making processes. The ethical basis of moral cosmopolitanism—a concern for the equally just and dignified treatment of all human life, or agential instances thereof—is leveraged, I argue, in a modest, incremental way that avoids usurping sovereignty. The first pillar adds practical substance to claims that effective information disclosure and utilisation are prerequisites for environmental dialogue (Eckersley, 2004: 116; Torgerson, 1999). When humans are denied the knowledge needed to enter into decision-making dialogue, they suffer imposed ignorance given their inability to consent to potentially harmful decision-making outcomes (Linklater, 2011: 99; Shapcott, 2008; Weaver, 2018). Attempting to overcome imposed ignorance, the pillar seeks to facilitate effective public participation in such processes; it seeks to set favourable conditions for state-public co-creation; and it thus contributes to an attempted enhancement of public trust in state-centric governance processes. With greater trust should come enhanced public perceptions of accountability (Abels, 2007; Biermann & Gupta, 2011; Keohane, 2011).

Such modest progress is far from being revolutionist. Any ethical progress is empowering rather than emancipatory. Accountability-holding remains itself a troublesome phenomenon, implicated as it is in power asymmetry. After all, to *attempt to* hold an actor accountable is to remain the weaker of the two actors in that relationship. To *attempt to* hold an actor accountable is to play the role of the *less* powerful accountability-holder, scrutinising the *more* powerful actor that is being held accountable. The less powerful accountability-holder suffers the heavier burden of effort, for they have the more arduous tasks of attempting to hold institutionalised power and authority to account, and of attempting to substantiate any improprieties alleged. These tasks are not insignificant. In fact, they are sometimes so dangerous that Aarhus' MOP recently had to initiate a rapid response mechanism, in order to protect environmental defenders exercising these and other Aarhus Convention rights (see this book, Chapter 5).

As such, the prerogatives of sovereignty never dissipate in the first pillar. One needs only ponder the disclosure exemptions to understand the persistence of *raison d'état*. Any solidarist progress is contained by, and confined within, existing power arrangements. There is no bold leap into unmanaged "open society" arrangements (see Reischauer & Ringel, 2023), themselves arguably destabilising of international order. Aarhus' Parties retain control over the conditions in which they permit or deny information disclosure. States' ultimate prerogatives—to protect their territories, populations and interests—are not hindered by the pillar's provisions.

Yet some counterbalancing against unbridled state predominance is observable. States' capacity to arbitrarily deny information was counterbalanced, inter alia, by the public interest test and the need to extract disclosable information from exempted material. Such counterbalancing sought to prevent a tyranny of sovereignty: it sought to avoid "carte blanche" administrative cultures of impunity.

This idea of counterbalancing, between human and sovereign prerogatives, is prescient for an English School treatment of solidarisation. State and public materialise as mutually regulating. We witnessed, here, germination of ambitious information access rights, challenging states to adopt a co-creative approach to governance. Yet they were counterbalanced by sovereignty. The rights never "cross the line" between what is palatable and unpalatable for Parties. The exemptions were *necessary*, in order for the first pillar to be regarded as a legitimate contribution to the ethical maturation of sovereignty. It would have been incomprehensible and impossible, had states not written the above exemptions into the first pillar. As such, whilst the pillar agitates states' ethical boundaries, challenging them to manoeuvre towards a more human-oriented "right to know", it remains itself delimited by the realities of sovereignty and the threshold of states' tolerance in a world where national interests and national security still matter.

The Second Pillar: Public Participation

It will be little surprise to the reader that the second pillar, public participation, further contributes to the Convention's operationalisation of the all-affected principle. The cosmopolitan impulse of human empowerment and co-creation is readily apparent. Before exploring it in earnest, let me sketch it here. Article 6 concerns public participation on specific activities, including those relating to GMOs. The latter were subject to a Convention amendment, adopted at the Second MOP (UNECE 2005). As I write, this "GMO Amendment" has not entered into force, requiring one final ratification (UNECE, 2023). Article 6 offers the most empowering participatory opportunities, and is enforced by Article 9.2 (Aarhus Convention, 1998; see this chapter, below). Article 7 provides participatory rights relating to environmental plans, programmes and policies, whilst Article 8 addresses public participation in law-making, or formally the "preparation of executive regulations and generally applicable legally binding normative instruments". Article 6 is most oriented towards co-creation; the latter articles provide vaguer rights. In this regard, the second pillar is characterised by a *gradient of participation* (Jendrośka, 2005: 16), starting with encouraging possibilities in Article 6, whose specific environment-related activities are most likely to affect humans and their environmental locales, whilst Articles 7 and 8 offer slimmer possibilities. Undue criticism is not intended: it is understandable that the public will, and should, be most able and willing to care for their "little patch" of environment, and hence empowered to influence decision-making on specific activities that might affect it (and them). The logic of this gradient coheres with Dryzek's (2000: 157) claim that those humans "in day-to-day contact with particular aspects of the ecosystem" are better placed "than distant managers or politicians to hear news from it" and to act accordingly. It also gels with what environmental communication scholars termed *self-in-place*, the sense of identification, emotional attachment and hence responsibility that humans feel vis-à-vis the environmental locales they occupy (Cantrill, 1998, 2011; Cantrill & Senecah, 2001). As will become more apparent below, Articles 7 and 8 deal with the more abstract and strategic govern*ment* of environmental decision-making, whilst Article 6 tackles more place-based, immediately impactful environment-related activities that are most likely to disrupt, undermine and potentially harm the environmental self-in-place. Neither is this gradient of public participation wildly incongruent with Sharpston's (2022: 120) finding that national-level organisations, responsible for strategic decision-making, surely have "less raw passion about the need to address a particular local problem than a concerned group of local citizens". I will now address the participatory provisions in turn.

Article 6, governing public participation vis-à-vis specific environmental activities, is the longest, most detailed and most precise provision in the second pillar (Lee & Abbot, 2003: 99; Oliver, 2013: 1440). And with good reason: the specific activities, enclosed in Annex I (Aarhus Convention, 1998), are most likely to have measurable direct impacts on specific environments occupied by humans. Specific activities may relate, inter alia, to energy and power, metals and minerals, chemicals, waste, timber, transport and water. They often relate to individual permit applications, by operators and developers, for very precisely defined projects or activities, in highly specified locations, and in strictly prescribed conditions (Jendrośka, 2012: 85). If an activity is not listed, but is subject to participatory provisions under domestic EIA law, then it is designated as such for Article 6 purposes (Aarhus Convention, 1998, Annex I.20). Similarly, Article 6.1b (Aarhus Convention, 1998) provides participatory rights in the context of decisions on activities not specified in Annex I, which "may have a significant effect on the environment". Such specific activities are not as abstract and politico-strategic as the subjects of decision-making in Articles 7 and 8. These activities yield grassroots impacts in clearly defined environmental locales, occupied by clearly identifiable human constituencies. Such activities are literally and figuratively "closer" to the humans they might harm.

It is in this context that the "main thrust" of Article 6 is towards the public concerned (Jendrośka, 2011: 124; see Aarhus Convention, 1998, Articles 6.2, 6.5, 6.6). After all, such specific activities are of risk and concern to "the public affected or likely to be affected by, or having an interest in, the environmental decision-making" (Aarhus Convention, 1998, Article 2.5). The broader "public" receives rights to submit comments (Aarhus Convention, 1998, Article 6.7) and be informed of the decision (Aarhus Convention, 1998, Article 6.9), but the public concerned is the chief beneficiary of Article 6 rights.

With this in mind, Article 6.2 requires the public concerned to "be informed, either by public notice or individually as appropriate, early in an environmental decision-making procedure, and in an adequate, timely and effective manner", of various factors relating to the activity. These include, inter alia, the nature of the application; the nature of possible decisions; the authority responsible for the decision; the participation procedure; and whether or not the activity is subject to EIA. Human empowerment surfaces in the requirement for early participation and for the adequacy, timeliness and effectiveness with which the public is informed. Reasonable timeframes for public participation are provided in Article 6.3 (Aarhus Convention, 1998), bolstered by Article 6.4's (Aarhus Convention, 1998) requirement for "early" public participation "when all options are open and effective public participation can take place".

None of this rich public participation in the permitting of specific activities can happen without accurate, current, fulsome environmental information. As such, the public concerned is entitled "access for examination, upon request where so required under national law, free of charge and as soon as it becomes available, to all information relevant to the decision-making…that is available at the time of the public participation procedure", without prejudicing the earlier disclosure exemptions (Aarhus Convention, 1998, Article 6.6; see this chapter, pillar one above). This provision constitutes recognition of the need for full and accurate knowledge, without which potentially affected persons would be caught in imposed ignorance. The logic is that state-human dialogue and co-creation can only be realised if potentially affected humans receive the "full facts". If humans and their associations only possess partial, limited knowledge, any dialogue and accountability-holding will remain plagued by the power asymmetries for which I was concerned earlier.

The cosmopolitan impulse of human empowerment is strengthened by Article 6.7's (Aarhus Convention, 1998) provision for the broader public "to submit, in writing or, as appropriate, at a public hearing or inquiry with the applicant, any comments, information, analyses or opinions that it considers relevant to the proposed activity". Crucially, "due account" needs to be taken, in the final decision, of the public participation that ensued (Aarhus Convention, 1998, Article 6.8), with prompt, written public disclosure of the decision alongside its "reasons and considerations" (Aarhus Convention, 1998, Article 6.9). "Due account" is a vital procedural safeguard, helping to ensure the worth of public contributions, and helping prevent token gesturing towards the public, under a veil of consultation.

I am not suggesting, of course, that accountable actors will change their proposals, in all circumstances, according to public opinion. This would be impossible, given the diversity of public views. It would also be undesirable, given the risk that value diversity would be homogenised and silenced by "lowest common denominator" solutions. Such "watering down" of public contributions would be counterintuitive to, and hindering of, state-human dialogue. But what I *would* expect, from this "due account" provision, is clear exposition of public contributions; explanation of where, and the degree to which, those contributions were integrated in the decision; and justification of where and why those contributions were not integrated in the ultimate decision. This again speaks to the all-affected principle, according to which, "what touches all should be heard and approved by all". This is because accountable actors need to account not only for the extent of their *reception* of public voice, but also of their *inclusion* and *exclusion* thereof.

However, some checks and balances serve to confine this moral cosmopolitan solidarism within Parties' thresholds of tolerance. Under Article 6.5 (Aarhus Convention, 1998, Article 6.9), Parties "should, where appropriate, encourage prospective applicants to identify the public concerned, to enter into discussions, and to provide information regarding the objectives of their application before applying for a permit". "Should" does not carry the legal compulsion of "shall". Such "identification of the public concerned" is worrisome (van Gool, 2022: 342; Weaver, 2015, 2018). To encourage applicants to "identify" the public concerned is to take the risk that applicants wilfully select those participants who are sympathetic to proposed decisions. This would reinforce the earlier power asymmetry in favour of the power*ful*. Another risk is that accountable actors pay token attention to public concern, in order to satisfy legal requirements. Such tokenism would involve orchestrating relevant participatory procedures and feigning a spirit of partnership, without fulsomely and sincerely taking due account of public contributions. This would also constitute an injustice, as potentially affected persons may well believe that their concerns are "heard", when in fact a foregone (and harmful) decision is masked by "consultation".

These risks speak to deeper issues of human dignity. If Article 6 (as with human rights more broadly) is applied with sincerity and humility by states, accountable actors will show true regard for the opinions, knowledge and concerns of the potentially affected. To apply these rights with sincerity and humility, in the spirit of the law, is to live out the trinity of voice: of access, standing and influence. It is to listen *for*, as well as to listen *to*, public voice (Dobson, 2014). This willingness to listen and engage in dialogue, remaining open to state-human co-creation, entails an investment of value and esteem in the human capacity to "hear" and "be heard", irrespective of wealth, status, political prowess or the "loudness" with which any actor articulates itself. This coheres with the cosmopolitan dialogue principle: that all actors engaging in dialogue enter it on as equal a footing as possible, with as equal a capacity as possible, and in the knowledge that the most convincing argument, irrespective of its source, will stand the greatest chance of taking precedence over the dialogue (Duffield Hamilton & Wills-Toker, 2006; Linklater, 2005).

Another check on moral cosmopolitan solidarism, in Article 6, is a separate provision that affords state discretion for national security purposes. Under Article 6.1c (Aarhus Convention, 1998), Parties may "decide, on a case-by-case basis if so provided under national law, not to apply the provisions of this article to proposed activities serving national defence purposes, if that Party deems that such application would have an adverse effect on these purposes". As with the information disclosure exemptions (see this chapter, pillar one above), this is a safeguard, weighted in favour of sovereign prerogatives, in contexts where public participation might jeopardise defence. It is a counterbalance against unbridled human participation, ensuring Aarhus' palatability for the consenting Parties.

Article 6, in my view, thus offers rich participatory possibilities. But recalling my opening treatment of the gradient of participation, it will now become apparent that Articles 7 and 8 are vaguer in their participatory provisions. Article 7, governing public participation in plans, programmes and policies relating to the environment, comprises but one paragraph. Plans, programmes and policies include development strategies, sectoral strategies (for instance regarding transport, water or health) and environmental action plans, at all government levels (Ebbesson et al., 2014: 176). As I discussed earlier, these governmental approaches to planning are more abstract and detached from local notions of *self-in-place*. The palpable immediacy of the impact yielded by these government interventions, on local self-in-place, will be less than that of the specific activities in Article 6. With this in mind, let me substantiate my claim of vagueness by quoting Article 7 (Aarhus Convention, 1998; emphasis added):

> Each Party shall make *appropriate* practical and/or other provisions for the public to participate during the preparation of plans and programmes relating to the environment, within a transparent and fair framework,

> having provided the necessary information to the public. Within this framework, article 6, paragraphs 3, 4 and 8, shall be applied. *The public which may participate* shall be identified by the relevant public authority, taking into account the objectives of this Convention. *To the extent appropriate*, each Party shall *endeavour* to provide opportunities for public participation in the preparation of policies relating to the environment.

Plans and programmes receive "appropriate" participatory provisions. Reasonable timeframes, opportunities for early participation and due account (Aarhus Convention, 1998, Articles 6.3, 6.4, 6.8) apply, but it is the public authority that needs to identify the "public which may participate". "May" is permissive and selective, perpetuating the power asymmetry for which I was concerned above (Weaver, 2015, 2018). With regard to policy, Parties must merely "endeavour" "to the extent appropriate" to provide participatory opportunities. "Appropriate" surfaces on a second occasion. It is therefore quite telling that an expert once deemed Article 7 as applying "in more recommendatory form" (Wates, 2005: 6). Sovereignty, in Article 7, is sovereignty *over* plans, programmes and policies. It is a gatekeeping sovereignty: Parties will "endeavour", only "to the extent appropriate", to make participatory arrangements in this more nebulous politico-strategic domain. Meanwhile, and with reference to authorities' responsibility to identify the public which may participate, Ebbesson et al. (2014: 179) seem rather keen indeed to remind practitioners that "the obligation...should not be seen as a tool to limit participation, but rather as a way to streamline the participation in order to make it more effective". Yet streamlining itself is implicated in power asymmetry: between the streamlining actor and those being streamlined. There is a risk that, for efficiency's sake, some "voices" are "streamlined out". This would present irreconcilable dilemmas for the sovereign gatekeepers, responsible for discerning those worthy of participation from those marginalised from it.

The gradient of participation declines more acutely with Article 8, on public participation during law-making, or formally the "preparation of executive regulations and legally binding normative instruments". Under Article 8 (Aarhus Convention, 1998; emphasis added), "Each Party shall *strive to promote* effective public participation at an appropriate stage, and while options are still open, during the preparation by public authorities of executive regulations and other generally applicable legally binding rules that may have a significant effect on the environment". Three steps "*should* be taken" to this end (Aarhus Convention, 1998, Article 8; emphasis added): (a) setting sufficient timeframes for public participation; (b) making draft rules publicly available; and (c) giving the public the opportunity to comment, directly or through representative consultative bodies. Tellingly, the final provision (Aarhus Convention, 1998, Article 8; emphasis added) states that the "result of the public participation shall be taken into account *as far as possible*". This is less consistent with the firmer cosmopolitan impulse of human empowerment and co-creation driving Article 6. Despite the use of the legally binding "shall" in "shall strive to promote", it has been noted that "the entire obligation is designed in a rather weak form" (Jendrośka, 2011: 96). "Striving to promote" effective public participation is tantamount, in my mind, to *working towards encouragement* of public participation: it is decidedly reticent. Meanwhile, the fact that public participation shall be taken account of "as far as possible" is rooted in sovereign pragmatism, implying that due account is limited in its possibilities in the politico-strategic domain of law-making. Mine is not an unfairly critical judgement, given that even the official implementation guide (Ebbesson et al., 2014: 181) deems Article 8 as evincing "a comparatively soft obligation to use best efforts, and uses indicative rather than mandatory wording for the steps to be taken".

Having considered the above, the following conclusions are drawn from the second pillar. Firstly, Aarhus' public participation provisions go a long way to breathing legal life into the cosmopolitan all-affected principle, dictating that "what touches all should be heard and approved by all". A cosmopolitan impulse centres, in the second pillar, upon the need for human empowerment, amplification of human voice and state-public co-creation. It is founded upon an assumption that the public will be most able and willing to influence decision-making procedures (a) that affect them and (b) when they are equipped with the most suitable procedural "tools" to do so. Those "tools"—the conditions needed in order to foster as effective a state-public partnership as possible—must optimise humans' capacity not only to *express* their voice but also to have it "*heard*" and *accommodated*. Human environmental voice must be credible and reasoned, irrespective of whether participants are laypersons or experts. I made this point earlier, when espousing a moral cosmopolitanism that counts one life, or agential instance thereof, as unconditionally and nonnegotiably equal as any other. And I reaffirm the point here, noting the solidarist potentialities of the second pillar whilst also remaining cognizant of the risks of residual power asymmetries. Any attempt by state Parties to arrange public participation in some form of hierarchy, discerning power*ful* from power*less* actors, will serve to perpetuate the harmful power asymmetries that Aarhus is designed to attenuate. It is, therefore, essential that Parties enter into the "spirit of the law", and deem even the vaguer participatory provisions as opportune springboards from which to institutionalise human empowerment. Both states and humans will be most acutely inclined towards co-creation in the context of locally relevant, place-based decision-making on specific activities, for it is in this domain that the sense of self-in-place is strongest, and the relation of vulnerability between potentially affected persons and proposed decisions the closest. But a necessary focus on specific environment-related activities should neither silence nor "crowd out" the participatory opportunities afforded in the more abstract and politico-strategic domains.

In this regard, secondly, we cannot deny the presence of a gradient of participation in the pillar, with the richest opportunities codified vis-à-vis specific activities and the vaguest emerging in the context of those latter, more abstract domains. This gradient has purchase for solidarisation. The more abstract and politico-strategic the decision-making domain, the less robust the participatory possibilities. Such logic constitutes a safeguard in favour of Parties' sovereign prerogatives. Whilst it is in the state and public interest for potentially affected persons to contribute their knowledge, expertise and concern to very specific instances of decision-making that might harm them and their environments, it is surely less palatable for states to co-create decisions in the more "high political" realm, which traditionally falls firmly within the purview of the state. Such counterbalancing between (a) human empowerment and (b) sovereign gatekeeping amplifies public voice in a controlled way, regulating it more in matters of more sensitive political weight. The sovereign prerogative, after all, is to protect the national interest, which will often animate the high political domain more than in the grassroots locale.

Such protection of sovereignty, thirdly, surfaces in the pillar's counterbalancing against "runaway" cosmopolitan dialogue. One check is the provision for the identification of the public concerned. There, risks of tokenism, selectiveness and the adoption of a mask of consultation were raised. Another check was the presence of state discretion, not least vis-à-vis Parties' consideration of national security in providing opportunities for public participation. As with the information disclosure exemptions (see this chapter, pillar one above), these are safeguards with which Parties seek to maintain control over their domestic affairs, thereby preserving the sanctity of their sovereignty. In this regard, the extent of public participation remains contained by, and confined within, states' thresholds of toleration. It does not undermine states' ability to preserve their interests. This echoes the book's chief finding: that Aarhus' solidarist, "humankind" progress is checked by sustained iterations of sovereignty. Mutual counterbalancing between human and sovereign prerogatives helps to achieve a rationalising human *empowerment* that avoids crossing into the realm of revolutionist *emancipation*.

The Third Pillar: Access to Justice

The third pillar, access to justice, is found in Article 9 (Aarhus Convention, 1998). It is a legal safeguard for the two preceding pillars, and is crucial to securing the legitimacy of state-public partnership and dialogue. I make this latter point because, as I argued earlier, there is no utility in environmental accountability-holding unless those persons exercising their rights can do so with the legal "teeth" of justiciability. This is what sets Aarhus apart from much of the legal and diplomatic context covered earlier (see this book, Chapter 3): much of that antecedent canon was aspirational and somewhat devoid of enforceability. The third pillar helps Aarhus go some way to enforcing procedural environmental rights.

To briefly map the third pillar, Article 9.1 (Aarhus Convention, 1998) assigns access to review procedures relating to Article 4 information requests. The procedures are available to any person who made an environmental request, in line with the "any person" right in Article 4. Article 9.2 (Aarhus Convention, 1998) assigns access to review procedures to the public concerned vis-à-vis (a) infringements of Article 6 participatory provisions for specific activities (themselves provided to the public concerned) and (b) other relevant provisions. Article 9.3 (Aarhus Convention, 1998) mandates broad public access to review procedures relating to alleged contraventions of national environmental law. This latter provision is the closest that the Aarhus Convention comes to establishing environmental *actio popularis*, given its empowerment of the public *en masse* to challenge environmental illegalities in Parties' jurisdictions. Meanwhile, Article 9.4 (Aarhus Convention, 1998) establishes minimum standards for review procedures, decisions and remedies, whilst Article 9.5 requires Parties to ensure effective access to justice.

Article 9.1 (Aarhus Convention, 1998; emphasis added) relates to Parties' compliance with their Article 4 access to information rights:

> "Each Party shall, within the framework of its national legislation, ensure that *any person* who considers that his or her request for information under article 4 has been ignored, wrongfully refused, whether in part or in full, inadequately answered, or otherwise not dealt with in accordance with the provisions of that article, has access to a review procedure before a court of law or another independent and impartial body established by law".

As such, an applicant may well have received some information, yet still have grounds for requesting a review procedure. "Independent and impartial bodies" need not be courts, but they should at the very least be independent, free from the political chains of government, bound by guarantees for due process and free from the regulatory constraints of private entities (Ebbesson et al., 2014: 191). Such bodies also need the legal competence to be able to make binding decisions on the public authority in question: nonbinding advisories would be insufficient, in this regard. Such safeguards are also provided in the need for expeditious, free or inexpensive review procedures (Aarhus Convention, 1998, Article 9.1). Final written decisions are binding on the public authority that holds the requested information, with reasons needing to be stated in writing (Aarhus Convention, 1998, Article 9.1).

Meanwhile, Article 9.2 (Aarhus Convention, 1998) relates primarily to Article 6 public participation provisions which, as we saw, apply to the public concerned. It obliges Parties to ensure that, within their national legal framework, the public concerned—where they have a sufficient interest, or where they maintain impairment of a right, in cases where this is a legal precondition—have access to a review procedure "before a court of law and/or another independent and impartial body established by law, to challenge the substantive and procedural legality of any decision, act or omission" under Article 6, and under any "other relevant provisions of this Convention" where provided under national law (Aarhus Convention, 1998, Article 9.2). This latter clause on "other relevant provisions" enables state Parties to extend, in their domestic law, Article 9.2's review procedures to other Aarhus Convention provisions, at their discretion. It thereby opens the possibility for state consent to be leveraged for more expansive and ambitious purposes.

NGOs meeting the definition of "public concerned" will meet the requirement, here, of having a sufficient interest or maintaining impairment of a right. It should also be noted that determination of whether an NGO indeed has a sufficient interest, or is in the position to maintain the impairment of a right, must be in compliance with Parties' national law and "consistently with the objective of giving the public concerned *wide access to justice*" (Aarhus Convention, 1998, Article 9.2; emphasis added). The latter italicisation denotes here that state Parties cannot, when applying the third pillar, use their discretion to limit persons' locus standi. The rhetoric of the official implementation guide is particularly instructive, in this regard, because it twice affirms that nothing prevents state Parties from granting access to justice to "any person without distinction" (Ebbesson et al., 2014: 194–195).

We should also recall, from the preceding section on the second pillar (see this chapter, above), that the broader "public" (as opposed to the public concerned) receive participatory rights in Articles 6.7 (on submitting comments etc.) and 6.9 (on being promptly informed of decisions) (Aarhus Convention, 1998). It is in this context that the broader public would also have access to justice under Article 9.2. Article 9.2 thus helps cultivate the territory on which the human individual can gain some degree of legal actorhood, in the domain of environmental decision-making, in their own right.

Article 9.3 (Aarhus Convention, 1998; emphasis added), furthermore, draws Aarhus as close as possible towards an environmental *actio popularis*, affording the general public, where they meet criteria specified in Parties' law, "access to administrative or judicial procedures to challenge acts and omissions by *private persons and public authorities*" contravening Parties' domestic law relating to the environment. This is potentially very highly empowering. It arguably goes some gratifying way towards setting a duty for state Parties to "facilitate environmental law enforcement by the public" (Hedemann-Robinson, 2022: 177). Such law enforcement may be invoked by the public directly or indirectly: *directly* through litigation in court, in circumstances other than to redress harm sustained by the complainant themselves; and *indirectly* by initiating review procedures in order to have relevant law enforced (Ebbesson et al., 2014: 197). The public would be within their rights to litigate against owners and operators of facilities implicated in environmental illegality, as well as public authorities themselves. Two factors are particularly auspicious for the cosmopolitan impulse of human empowerment. Firstly, an allegation of illegality is enough to trigger a review procedure under this provision (Ebbesson et al., 2014: 197). Secondly, the law being invoked must merely *relate* to the environment: it need not be a discretely environmental law. This theoretically sets a favourable context in which a broad public would be able to initiate legal review with regard to an equally broad range of alleged illegalities. The provision is potentially pioneering in empowering humans' legal actorhood.

The cosmopolitan impulse is practically reinforced by the requirement for "adequate and effective remedies, including injunctive relief as appropriate" that are "fair, equitable, timely and not prohibitively expensive" (Aarhus Convention, 1998, Article 9.4). Resultant decisions are to be written and publicly accessible (Aarhus Convention, 1998, Article 9.4). Such practical reinforcement is further bolstered by the requirement for Parties to ensure that information is provided to the public on the above review procedures, and also by the need for assistance mechanisms to be established, so as to remove or reduce financial or other barriers to access to justice (Aarhus Convention, 1998, Article 9.5). This conjures, in my mind at least, such assistance mechanisms as environmental legal aid, and also environmental awareness and education: not only for human individuals and NGOs, but also for public and private sector actors. This resonates with my earlier call for an environmental widening participation programme. "Adequacy and effectiveness", furthermore, denote the need for review procedures' practical capacity to ensure palpable access to justice. The provision for injunctive relief is also theoretically robust, given that injunctive relief seeks to stop or prevent unlawful actions from being conducted, and thus should incur a cessation of the harm being sustained. One might understand injunctive relief, in environmental decision-making, as a green take upon the "cease and desist" device in law. But environmental injunctive relief is replete with potential risks and insufficiencies. Elsewhere (Weaver, 2015, 2018), I have discussed the risks of injunctive relief yielding only pyrrhic victories: instances of only formal progress towards cessation of harm, which come too little, too late, and often *after* the incidence of irreversible harm (Khalastchi, 1999: 304; Marsden, 2002: 36; Nordberg, 2007: 95; Working Group on Access to Environmental Justice, 2008).

It is therefore crucial that Parties apply Aarhus' access to justice provisions swiftly and meticulously, in order for *procedural* rights to truly serve the public in its quest to prevent *substantive* environmental harm. This spirit of procedural empowerment is fortified by the provision for "Fairness, equity and timeliness", which seeks a justice that can be accessed expeditiously by all persons, irrespective of their status, privilege and influence. The need for review to not be prohibitively expensive is designed to optimise the public's chances to achieve actual justice. The journey towards breaking the barriers of prohibitive expense was not the easiest, although some modicum of progress was made in recent years (Holder & Lee, 2007: 118; Oliver, 2013; Pedersen, 2011).

Bearing in mind the foregoing, the crux is that environmental agreements are likely to remain aspirational unless they are endowed with the legal "teeth" and "bite" of justiciability. Unless environmental rights and duties are codified in legally binding instruments that are deployed in courts or other sufficiently competent judicial sites, they will remain the hopes of the environmentally virtuous. To overcome these risks of aspiration, the third pillar writes into Aarhus' operative text the conditions in which the public can seek legal review, not only of first and second pillar contraventions, but *also*—more ambitiously—of infringements of domestic law relating to the environment. This constitutes a "sharpening" of Aarhus' legal "teeth", especially in the context of the latter push towards environmental *actio popularis*, which has the proclivity to intrude quite considerably upon Parties' sovereignty. It opens Parties to potentially broad public scrutiny, not only relating to their discrete Aarhus compliance, but also to their far broader compliance with their corpus of law relating to the environment. Sovereignty is agitated in this regard, because it has to work "harder"—more responsibly, more judiciously, more compliantly—in order to retain its legitimacy. Parties must meet stricter standards of civilisation (Gong, 1984) in order to retain their sovereign privileges in the international society arrangements delineated by Aarhus Party membership. They must protect a more expansive form of human legal actorhood, in environmental contexts.

Yet such a more demanding approach to sovereignty is counterbalanced by the fact that the ultimate responsibility for determining locus standi continues to rest with Aarhus' state Parties (Hedemann-Robinson, 2022: 177). We must also acknowledge that the very architectures of legal review—those courts and other bodies with sufficient legal competence—are embedded irremovably from the terrain of state sovereignty. It is from that terrain that they derive their authority to function. What needs to be recognised is that whilst these third pillar rights have contributed to a reform in environmental justice practices, they remain *bestowed upon* humans and their associations. In the relation of bestowal, between the rights-holder and rights-provider, there continues to be a residual power asymmetry. And where justice is sought—and swiftly so, in the case of injunctive relief—it may still be pyrrhic in its nature.

Conclusion

If the preceding chapter (Chapter 3) traced Aarhus' *propagation*, and if the next chapter (Chapter 5) explores its *growth*, this chapter addressed Aarhus' *germination*, by means of its legally binding trinity of procedural environmental rights contained in the operative text. The lesson is that any solidarist progress, sought in an international environmental agreement, will be delimited by consenting states' thresholds of tolerance. Any human-oriented solidarist progress will remain as radical as consenting states permit. It will be accompanied by (a) states' reaffirmations of their sovereignty and (b) attempts to prevent the undermining of sovereignty, through counterbalancing in favour of sovereign prerogatives. Such counterbalancing seeks to solidify states' control over their own destinies: over their power as the world's sovereigns; over the legitimacy with which they keep control of their territories, populations, national interests and security priorities. Such counterbalancing in favour of sovereignty reduces the risk of usurpation, thereby attenuating the risk of international order being destabilised.

It is in this context that I observed evolutionary solidarist progress in Aarhus' procedural trinity. Such progress agitates sovereignty, whilst itself being delimited by the power frameworks in which the agitation occurs. This is indicative of solidarisation, in demonstrating the mutual reinforcement of solidarist endeavours and existing pluralist arrangements. Of course, I do not wish to downplay the important progress achieved in the trinity. The latter was seen to contribute to an amplification of human environmental voice; a codification of concern for state-human partnership and dialogue; and a garnering of co-creation in a more dialogic and responsible form of environmental decision-making.

Valuable and innovative tools, which were elicited from the trinity, help to achieve this. They included, inter alia, a turn from "need to know" to "right to know" information governance; an operationalisation of the all-affected principle; and some degree of legal codification of that principle, not least via the "due account" provision. Checks on the risks of sovereign tyranny were also seen to be employed in the procedural trinity. It is worth recalling, here, the public interest test in information disclosure, alongside the need to extract disclosable information from otherwise exempted context. These tools seek to prevent states from feigning interest in, or abusing power asymmetries vis-à-vis, human agency. The trinity's provisions seek to entrench a sincerity and humility in the Parties implementing their Aarhus duties. The risk of the violence of silence (Brisman, 2013) being institutionalised is further countered by a concerted effort, in the third pillar, to set the conditions in which humans and their associations can benefit from legal actorhood. Evidence of a "firming up" of justiciability was gauged, not least from Aarhus' push towards environmental *actio popularis*.

Of course, I am not blind to the challenges of entrenching the cosmopolitan impulse that drives these provisions. Any moral cosmopolitanism entrenched, here, is cautiously incremental in its modest empowerment of humans and their associations, and its steering well clear of emancipatory rhetoric and ambition. Nowhere was sovereignty undermined, in the above account. The legitimacy with which states pursue their interests and security priorities is safeguarded in the exemptions that were discussed. Similarly, states' capacity to identify the public concerned, to retain sovereignty over more abstract "high political" decision-making and to determine locus standi are indicative of the persistence of, and insistence upon, sovereignty. Such is a counterbalancing against unbridled revolutionism.

Any solidarist ethical progress is thus incremental, necessarily modest and delimited by sovereignty. Very slim progress was made towards Linklater's (1998: 7) notion of "institutional frameworks which widen the boundaries of the dialogic community", which enable "radical encroachments" on sovereignty and which demonstrate the feasibility of initiating "transfers of state power to sub-national and transnational political authorities". The cosmopolitan essence of Linklater's transformative vision is captured in the procedural trinity. The essence of the transformative vision is distilled, and infuses the existing, orthodox, risk averse and order-oriented power frameworks of environmental multilateralism. Such infusion of transformative potentialities into those frameworks dilutes the potency of the attempted transformation, such that the potency remains only as intense as sovereignty allows.

Note

1. For brevity, in this chapter, I will make reference within the text to "Aarhus Convention 1998".

Bibliography

Abels, G. (2007). Citizen involvement in public policymaking: Does it improve democratic legitimacy and accountability? The case of pTA. *Interdisciplinary Information Sciences, 13*(1), 103–116.

Ahrens, B. (2017). *The solidarisation of international society: The EU in the global climate change regime* (GLOBUS Research Paper 5/2017). University of Oslo.

Ahrens, B. (2018). *AmbigEUity: The EU and the solidarisation of international society* (PhD thesis). Eberhard Karls Universität Tübingen.

Ahrens, B. (2019). The European Union between solidarist change and pluralist re-enactment. In T. Brems Knudsen & C. Navari (Eds.), *International organisation in the anarchical society: The institutional structure of world order* (pp. 265–292). Palgrave Macmillan.

Ahrens, B., & Diez, T. (2015). Solidarisation and its limits: The EU and the transformation of international society. *Global Discourse, 5*(3), 341–355.

Arnstein, S. (1969). A ladder of participation. *Journal of the American Institute of Planners, 35*(4), 216–224.

Ball, T. (2006). Democracy. In A. Dobson & R. Eckersley (Eds.), *Political theory and the ecological challenge* (pp. 131–147). Cambridge University Press.

Biermann, F., & Gupta, A. (2011). Accountability and legitimacy in earth system governance: A research framework. *Ecological Economics, 70*, 1856–1864.

Blundell, D. (2015). Access to and collection of environmental information. In C. Banner (Ed.), *The Aarhus Convention: A guide for UK Lawyers* (pp. 99–120). Hart.

Böhmelt, T., Koubi, V., & Bernauer, T. (2014). Civil society participation in global governance: Insights from climate politics. *European Journal of Political Research, 53*, 18–36.

Brisman, A. (2013). The violence of silence: Some reflections on access to information, public participation in decision-making, and access to justice in matters concerning the environment. *Crime, Law and Social Change, 59*, 291–303.

Cabinet Office. (2018). *Government security classifications*. https://assets.publishing.service.gov.uk/government/uploads/system/uploads/attachment_data/file/715778/May-2018_Government-Security-Classifications-2.pdf. Accessed 6 May 2023.

Cantrill, J. (1998). The environmental self and a sense of place: Communication foundations for regional ecosystem management. *Journal of Applied Communication Research, 26*(3), 301–318.

Cantrill, J. (2011). The role of a sense of self-in-place and risk amplification in promoting the conservation of wildlife. *Human Dimensions of Wildlife, 16*(2), 73–86.

Cantrill, J., & Senecah, S. (2001). Using the 'sense of self-in-place' construct in the context of environmental policy-making and landscape planning. *Environmental Science and Policy, 4*(4), 185–203.

Convention on Access to Information, Public Participation in Decision-making and Access to Justice in Environmental Matters. (1998). ECE/CEP/43. https://unece.org/DAM/env/pp/documents/cep43e.pdf. Accessed 9 March 2023.

Cox, R. (2006) *Environmental communication and the public sphere*. Sage.

Cramer, B. W. (2009). The human right to information, the environment and information about the environment: From the Universal Declaration to the Aarhus Convention. *Communication Law and Policy, 14*, 73–103.

Data Protection Act. (2018). https://www.legislation.gov.uk/ukpga/2018/12/contents/enacted. Accessed 18 June 2023.

Dobson, A. (2014). *Listening for democracy: Recognition, representation, reconciliation*. Oxford University Press.

Dryzek, J. (2000). *Deliberative democracy and beyond*. Oxford University Press.

Duffield Hamilton, J., & Wills-Toker, C. (2006). Reconceptualising dialogue in environmental public participation. *Policy Studies Journal, 34*(4), 755–775.

Ebbesson, J., Gaugitsch, H., Jendroska, J., Marshall, F., & Stec, S. (2014). *The Aarhus Convention: An implementation guide*. United Nations.

Eckersley, R. (2000). Deliberative democracy, representation and risk. In M. Saward (Ed.), *Democratic innovation*. Routledge.

Eckersley, R. (2004). *The green state: Rethinking democracy and sovereignty*. MIT Press.

Falkner, R. (2021). *Environmentalism and global international society*. Cambridge University Press.

Florini, A. (2007). *The right to know: Transparency for an open world*. Columbia University Press.

Gavouneli, M. (2000). Access to environmental information: Delimitation of a right. *Tulane Environmental Law Journal, 13*, 303–327.

Gilmore, J. (2023). Conceptualising good global statehood: Progressive foreign policy after the populist movement. *International Theory, 15*(1), 79–105.
Gong, G. (1984). *The standard of civilisation in international society*. Clarendon Press.
Hallo, R. (2011). Access to environmental information: The reciprocal influences of EU law and the Aarhus Convention. In M. Pallemaerts (Ed.), *The Aarhus Convention at ten: Interactions and tensions between conventional international law and EU environmental law* (pp. 55–66). Europa Law.
Harman-Stokes, K. M. (1995). Community right-to-know in the newly independent states of the former Soviet Union: Ending the culture of secrecy surrounding the environmental crisis. *Virginia Environmental Law Journal, 15*, 77–138.
Hedemann-Robinson, M. (2022, June). Access to environmental justice and European Union institutional compliance with the Aarhus Convention: A rather longer and more winding road than anticipated. *European Energy and Environmental Law Review*, 175–188.
Holder, J., & Lee, M. (2007). *Environmental protection, law and policy*. Cambridge University Press.
Home Office. (2023). *New espionage offences: Factsheet*. https://www.gov.uk/government/publications/national-security-bill-factsheets/espionage-etc-national-security-bill-factsheet. Accessed 6 May 2023.
Jendrośka, J. (2005). Aarhus Convention and community law: The interplay. *Journal for European Environmental and Planning Law, 1*, 12–21.
Jendrośka, J. (2011). Public participation in environmental decision-making: Interactions between the Convention and EU law and other key legal issues in its implementation in the light of the opinions of the Aarhus Convention Compliance Committee. In M. Pallemaerts (Ed.), *The Aarhus Convention at ten: interactions and tensions between conventional international law and EU environmental law* (pp. 91–148). Europa Law Publishing.
Jendrośka, J. (2012). Citizens' rights in European environmental law: Stocktaking of key challenges and current developments in relation to public access to information, participation and access to justice. *Journal for European Environmental and Planning Law, 9*(1), 71–90.
Keohane, R. (2011). Global governance and legitimacy. *Review of International Political Economy, 18*(1), 99–109.
Khalastchi, R. (1999). International environmental law in the courts of the United Kingdom. *Review of European Community and International Environmental Law, 8*(3), 301–308.
Kingdom', Review of European Community and International Environmental Law, 8(3), pp.301–308.

Klinke, A. (2012). Democratising regional environmental governance: Public deliberation and participation in transboundary ecoregions. *Global Environmental Politics, 12*(3), 79–99.

Krämer, L. (2012). Transnational access to environmental information. *Transnational Environmental Law, 1*(1), 95–104.

Lee, M., & Abbot, C. (2003). The usual suspects? Public participation under the Aarhus Convention. *The Modern Law Review, 66*(1), 80–108.

Linklater, A. (1998). *The transformation of political community*. Polity.

Linklater, A. (2002). Cosmopolitan political communities in international relations. *International Relations, 16*(1), 135–150.

Linklater, A. (2005). Dialogic politics and the civilising process. *Review of International Studies, 31*, 141–154.

Linklater, A. (2011). *The problem of harm in world politics: Theoretical investigations*. Cambridge University Press.

Marsden, S. (2002). Protecting archaeological heritage in Wetlands: The muddied waters of international. *European, English and Australian Law', Environmental Law Review, 4*, 26–50.

Mason, M. (2010). Information disclosure and environmental rights: The Aarhus Convention. *Global Environmental Politics, 10*(3), 10–31.

Nordberg, L. (2007). The Vuosaari Harbour case: Implementation of the habitats and birds directives in the Vuosaari Harbour project. *Review of European Community and International Environmental Law, 16*(1), 87–103.

Oliver, P. (2013). Access to information and to justice in EU environmental law: The Aarhus Convention. *Fordham International Law Journal, 36*, 1423–1470.

Pallemaerts, M. (2011). 'Introduction'. In M. Pallemaerts (Ed.), *The Aarhus Convention at ten: Interactions and tensions between conventional international law and EU environmental law* (pp. 1–16), Europa Law

Pedersen, O. W. (2011). Price and participation: The UK before the Aarhus Convention's compliance committee. *Environmental Law Review, 13*(2), 115–123.

Protocol on Pollutant Release and Transfer Registers to the Convention on Access to Information, Public Participation in Decision-Making and Access to Justice in Environmental Matters3.. (2003). United Nations Treaty Series, Volume 2629, A-37770. https://treaties.un.org/doc/Publication/UNTS/Volume%202629/volume-2629-A-37770.pdf. Accessed 7 May 2023.

Reed, M. S. (2008). Stakeholder participation for environmental management: A literature review. *Biological Conservation, 141*, 2417–2431.

Reischauer, G., & Ringel, L. (2023). Unmanaged transparency in a digital society: Swiss army knife or double-edged sword? *Organisation Studies, 44*(1), 77–104.

Richardson, B., & Razzaque, J. (2006). Public participation in environmental decision-making. In B. Richardson & S. Wood (Eds.), *Environmental law for sustainability* (pp. 165–194). Hart.

Sand, P. H. (2003). Information disclosure as an instrument of environmental governance. *Heidelberg Journal of International Law, 63*, 487–502.

Saward, M. (2006). Representation. In A. Dobson & R. Eckersley (Eds.), *Political theory and the ecological challenge* (pp. 183–199). Cambridge University Press.

Senecah, S. (2004). The trinity of voice: The role of practical theory in planning and evaluating the effectiveness of environmental participatory processes. In S. P. Depoe, J. W. Delicath, & M. A. Elsenbeer (Eds.), *Communication and public participation in environmental decision making* (pp. 13–33). SUNY Press.

Shapcott, R. (2008). Anti-cosmopolitanism, pluralism and the cosmopolitan harm principle. *Review of International Studies, 34*, 185–205.

Sharpston, E. (2022). From "do trees have rights?" To wondering about ecocide: Some legal reflections. *Environmental Policy and Law, 52*, 117–131.

Spyke, N. S. (1999). Public participation in environmental decisionmaking at the new millennium: Structuring new spheres of public influence. *Boston College Environmental Affairs Law Review, 26*(2), 263–313.

Steele, J. (2001). Participation and deliberation in environmental law: Exploring a problem-solving approach. *Oxford Journal of Legal Studies, 21*(3), 415–442.

Torgerson, D. (1999). *The promise of green politics: Environmentalism and the public sphere.* Duke University Press.

United Nations Economic Commission for Europe. (2005). Report of the Second Meeting of the parties: Decision II/1—Genetically modified organisms, adopted at the Second Meeting of the parties held in Almaty, Kazakhstan, on 25–27 May 2005. ECE/MP.PP/2005/2/Add.2. https://unece.org/DAM/env/documents/2005/pp/ece/ece.mp.pp.2005.2.add.2.e.pdf. Accessed 8 May 2023.

United Nations Economic Commission for Europe. (2023). GMO amendment. https://unece.org/environment-policy/public-participation/aarhus-convention/gmo-amendment. Accessed 8 May 2023.

Van Gool, E. (2022, October). Searching for 'environmental justice' in EU environmental law. *European Energy and Environmental Law Review, 31*(5), 334–346.

Warren, M. (2017). *The all-affected interests principle in democratic theory and practice* (Working Paper). Institute for Advanced Studies, Vienna. http://aei.pitt.edu/93142/1/Warren_-_All_Affected_Interests_Principle.pdf. Accessed 8 May 2023.

Wates, J. (2005). The Aarhus Convention: A driving force for environmental democracy. *Journal of European Environmental and Planning Law, 1*, 2–11.

Weaver, D. (2015). *The Aarhus Convention: Towards a cosmopolitan international environmental politics* (Unpublished PhD thesis). Keele University. https://eprints.keele.ac.uk/id/eprint/2310/1/Weaverphd2015.pdf. Accessed 12 April 2023.

Weaver, D. (2018). The Aarhus Convention and process cosmopolitanism. *International Environmental Agreements, 18*(2), 199–213.

Weaver, K. (2023). Bridging indigenous and western knowledge-systems in knowledge co-production with Amazonian indigenous communities: A systematic realist review. *Development Studies Research, 10*(1), 1–22.

Working Group on Access to Environmental Justice. (2008). Ensuring access to environmental justice in England and Wales. *The Sullivan Report*. http://assets.wwf.org.uk/downloads/justice_report_08.pdf. Accessed 11 May 2023.

Yanitsky, O. N. (2012). From nature protection to politics: The Russian environmental movement, 1960–2010. *Environmental Politics, 21*(6), 922–940.

CHAPTER 5

Growth: The Aarhus Convention's Organisational Infrastructure

Introduction

This book has observed a cosmopolitan impulse—a turn towards the "humankindness" of moral cosmopolitanism—driving Aarhus' solidarist ambitions. This impulse was seen to animate not only Aarhus' propagation from the postwar period onwards (see this book, Chapter 3), but also to animate its germination by means of its legally binding trinity of procedural environmental rights (see this book, Chapter 4). However, neither have we witnessed any manoeuvre towards unbridled human emancipation, nor a firmly consolidated turn to cosmopolitan governance per se. Throughout the book, the cosmopolitan impulse was seen to be contained by, and confined within, states' threshold of tolerance. The pioneering solidarist progress, accounted for in this book, was seen to occur only to the extent that it remained palatable for Aarhus' consenting states. This chief finding—a finding that resonates with the logic of solidarisation—is reinforced here, in a chapter that examines Aarhus' growth, namely in terms of the development of its organisational infrastructure.

I argue, in this chapter, that the amplification of human voice, and the institutionalisation of human empowerment, intensified as Aarhus "grew" following its germination. This is largely attributable to the precedent that was set during the Convention's preparations and negotiations. The broad and deep human participation, during propagation, set "a platform for civil society to influence its implementation and enforcement" (Day,

D. Weaver, *The Aarhus Convention*, Environmental Politics and Theory,
https://doi.org/10.1007/978-3-031-43536-2_5

2015: 184–185). In this regard, the chapter will observe, in the strategic, operational and tactical tiers of the infrastructure, a concern for amplifying human voice, and for foregrounding human inclusion in environmental multilateralism. The strongest potential is identified in the tactical tier, the compliance mechanism. There, humans and their associations are most keenly empowered to initiate actions that can incur domestic change inside Party jurisdictions. It is there that Aarhus pushes again towards environmental *actio popularis* (see also this book, Chapter 4, pillar three), enabling humans, NGOs and combinations thereof to directly lodge communications of Parties' alleged non-compliance, and thus to directly hold Parties accountable.

Emancipatory potential is not gauged from Aarhus' organisational infrastructure, in this chapter. A modest encroachment upon, and intrusion into, sovereignty is seen to coincide with a more rationalising approach to human empowerment. The "buck" of power continues to "stop" with states. Whilst glimpses of cosmopolitanism are elicited from the infrastructure, much counterbalancing in favour of sovereignty is evident. Indeed, on a couple of occasions, we will witness states' unilateral reticence vis-à-vis the Convention. I refer, here, not only to Belarus' reservations relating to the initiation of a rapid response mechanism and associated Special Rapporteur on Environmental Defenders, preceding Belarus' withdrawal from the Convention. I also refer to the US' refusal to recognise Aarhus' compliance mechanism. As we shall see, this latter unilateral reticence occurred at a time when the US refused to become Party to the international legal arrangements for the International Criminal Court (ICC). It is worth noting, at this early point, that the US' tone in both instances of its unilateralism is similar, in terms of its autonomous refusal to accept a legal instrument's legitimacy. Let me not recount the whole story here, but suffice it to note that we will witness multiple other iterations of sovereignty, and instances of sovereign counterbalancing.

What follows, then, is an observation of human-oriented solidarist ambition being delimited within states' thresholds of tolerance. Human voice is amplified, and human inclusion institutionalised, in the organisational infrastructure. This serves to ethically "stretch" Parties' thresholds of ethical ambition, demanding more from a sovereignty that becomes just a little more acutely attuned to human and environmental vulnerabilities. But such recalibration of sovereignty does not indicate post-sovereign transformation. Sovereignty concomitantly (a) is subjected to ethical maturation whilst (b) delimiting Aarhus' solidarist ambition, in order for it to remain palatable for consenting states. I will now substantiate this logic with reference to the strategic, operational and tactical tiers.

The Strategic Tier: The Meeting of the Parties

The MOP, termed by other international environmental agreements the Conference of the Parties (COP), is an agreement's "highest authority" (Andresen, 2007: 331), responsible for internal decision-making, codifying legal provisions, regulating compliance and instigating external relations (Churchill & Ulfstein, 2000: 626). Aarhus' MOP is established under Article 10 (Convention on Access to Information, Public Participation in Decision-making and Access to Justice in Environmental Matters, 1998). It reinforces the principle of sovereign equality by providing one vote to one Party, and by assigning to regional economic integration organisations "a number of votes equal to the number of their member States which are Parties to this Convention" (Convention on Access to Information, Public Participation in Decision-making and Access to Justice in Environmental Matters, 1998: Article 11). Such organisations "shall not exercise their right to vote if their member States exercise theirs, and vice versa" (Convention on Access to Information, Public Participation in Decision-making and Access to Justice in Environmental Matters, 1998: Article 11.2). An EU member state cannot, therefore, vote twice. This "one Party, one vote" arrangement is consistent with the traditions of international law, emphasising the parity of sovereign states. Votes are thus "not weighted and each Party has the same right to participate" (Ebbesson et al., 2014: 216).

With this in mind, a cardinal function of Aarhus' MOP is to keep implementation under "continuous review" (Convention on Access to Information, Public Participation in Decision-making and Access to Justice in Environmental Matters,1998: Article 10.2) through reviewing the Convention's legal provisions; facilitating internal/external cooperation; establishing subsidiary bodies, protocols and amendments; pursuing Aarhus' goals via additional actions; adopting rules of procedure and implementing Article 5.9 (Convention on Access to Information, Public Participation in Decision-making and Access to Justice in Environmental Matters, 1998), which was actioned in the PRTR Protocol (Protocol on Pollutant Release and Transfer Registers to the Convention on Access to Information, Public Participation in Decision-Making and Access to Justice in Environmental Matters, 2003). It should also come as no surprise that the MOP has financial duties (Convention on Access to Information, Public Participation in Decision-making and Access to Justice in Environmental Matters, 1998: Article 10.3). Taking all of these functions into consideration, we should agree that the MOP, the Convention's "high political" arm, is consistent with mainline multilateralism. And it is to be expected that, in such a strategic arm of the infrastructure, iterations of sovereignty will be palpable as states seek to retain and reassert their control over an instrument to which they voluntarily consent.

Such iterations of sovereignty are not unproblematic. In this regard, the next section explores a key tool for "continuous review", the requirement for Parties' national implementation reporting (NIR) arrangements. Afterwards, attention is paid to the rhetoric of the MOPs that have been held to date. Whilst glimpses of cosmopolitanism are gauged, scarce digression from orthodox multilateralism is apparent. A certain doggedness of state sovereignty, in fact, will be elicited from what follows.

The Doggedness of Sovereignty: National Implementation Reporting

A key tool for "continuous review" is the provision for Aarhus Parties' NIR arrangements (see UNECE [2023a] for a comprehensive searchable database). Parties are requested to provide update reports to the MOP, on the state of their Convention implementation, every three years (Department for Environment, Food and Rural Affairs, 2020). This helps to ensure strategic oversight of implementation, alongside any necessity for resultant interventions. NIR arrangements were mandated at the first MOP in 2002, in Decision I/8 (UNECE, 2004b). The latter required Parties to report on the "necessary legislative, regulatory or other measures" taken in a bid to implement the Convention (UNECE, 2004b: 2). Reports need to be prepared transparently and in public consultation (see Department for Environment, Food and Rural Affairs [2020] for evidence of this). This requirement for public participation in NIR was likely "unprecedented" in the erstwhile reporting mechanisms of international environmental agreements (Wates, 2005: 8). Signatories and other non-Party states can voluntarily make their own NIR submissions, as can IOs and also NGOs "engaged in programmes or activities providing support to Parties and/or other States" in implementation (UNECE, 2004b: 2). Decision I/8's guidance on formulating NIR submissions was augmented by further comprehensive guidance early in Aarhus' history (UNECE, 2007).

It is necessary to focus our sights on NIR, here, because the archives would indicate some difficulty with these arrangements. If anything, the state of NIR performance indicates the doggedness of state sovereignty. It is noteworthy that as early as the second MOP, Decision II/10 (UNECE, 2005d: 2) was compelled to note "with concern the failure by some Parties to submit their reports" and the failure by others to submit by the deadline. Decision II/10 (UNECE, 2005d: 2; emphasis added) "*urge[d]* all Parties to comply with decision I/8 and other stipulations developed under the reporting mechanism". Yet it also (UNECE, 2005d: 2; emphasis added) "*invite[d]* Parties...to provide more information with regard to the practical implementation of each of the Convention's provisions...and to indicate *any major differences of opinion* emerging from the consultation process". This indicates that by the second MOP, Parties were (a) providing insufficient detail in their NIR submissions; (b) insufficiently accommodating the diversity of human voice in their NIR public consultations and (c) "satisficing" (concurrently satisfying and sufficing) their "letter of the law" duties, and as such necessitating "*invitation*" to go further towards the spirit of the law, and provide more detail in their NIR submissions.

This was not a solidarist triumph. Most troubling of all was the risk of diversity being homogenised during public consultation. After all, the presence of "major differences of opinion" would surely indicate that public consultation (a) actually occurred and (b) was not silenced or subjected to homogenising tendencies. The absence of "major differences of opinion", during and after such public consultation, would, I fear, indicate the silencing, homogenisation or otherwise ignorance of diverse human voice (see Brisman, 2013). The latter would constitute an injustice in and of its own right, and would undermine the cosmopolitan and democratising project for which Aarhus was mobilised in the first place.

Whilst more than three quarters of Parties had submitted their NIR documents by the fourth MOP, many had failed to meet the deadline (UNECE, 2011b: 11). The MOP (UNECE, 2011b: 11; emphasis added) responded by "*Reiterat[ing] its encouragement* of Parties to start the preparation of national implementation reports...sufficiently in advance of, and at the latest five months before, the deadline...with a view to ensuring meaningful public consultation on the reports at the national level". By the sixth MOP (UNECE, 2017b: 10), one third of Parties had not submitted their reports by the deadline, and a diversity of states (Armenia, Bosnia and Herzegovina, The Netherlands, Moldova, Macedonia, Ukraine and the UK) had failed to submit by the time the report of the sixth MOP was issued.

As I complete this book, the report of the latest seventh MOP (UNECE, 2022b: 8) noted "with regret that 40 per cent of Parties (19 Parties) had not submitted their reports by the set deadline". Solemnly, the seventh MOP (UNECE, 2022b: 8–9; emphasis added) also expressed "*deep concern* that the Republic of Moldova had also still not yet submitted its national implementation report for the fifth reporting cycle – the only country that had not done so – and called upon the Compliance Committee…to consider the *ongoing failure* by the Republic of Moldova to submit its reports for the fifth and sixth cycles". This is very stern rhetoric. Of the Parties that *did*, furthermore, submit reports in this latest cycle, many "provided overwhelmingly detailed information on existing, amended and/or updated legislation…without providing information on its relevance to the implementation of relevant provisions of the Convention" (UNECE, 2021: 4). Some Parties additionally either failed to answer questions on implementation difficulties, repeated themselves, or used NGOs' information when formulating their NIR statements (UNECE, 2021: 5).

It is thus not wholly controversial to infer that Aarhus' NIR arrangements are experiencing ongoing difficulties at the time of writing. The above coverage provides a sound reminder that the diplomatic "pitch" is one on which states enter and withdraw at will. Allusions to "invitation" and "encouragement" underline the persistence of state sovereignty and state consent. But to momentarily adopt a vantage point of government, there is a certain pragmatism in the above account. After all, to make such a NIR submission is not only to showcase states' best practice. It is also to throw into stark relief, at the very least, Parties' difficulties discharging their responsibilities and, in worst case, to render themselves liable for litigation. Nowhere is it in the nature of sovereignty to wilfully incriminate itself. The retort from a statesperson, to continue briefly adopting this vantage point, would surely be "*better a late NIR submission than never*".

This underscores the tensions between the "letter of the law", an emblem of pluralist minimalism and sovereign prerogative, and the "spirit of the law", the driver of Aarhus' solidarist ambition. Through English

School lenses, it is the letter of the law that regulates and constrains the more ethically charged, maximalist endeavours of solidarism.

Meeting of the Parties' Rhetoric: 'Humankindness' Tempered by Sovereignty

It should be clear, from the preceding content, that tension exists between Parties' (difficulties with) adherence to the *letter* of the law and the more solidarist *spirit* of the law. Tension is also evident in the interplay between a rhetoric of "humankindness", elicited from the MOPs held to date, and the counterbalancing in favour of sovereignty that serves to temper it. Before I assess the rhetoric of the successive MOPs, and pay attention to their resultant declarations and outcomes, I should remark that a "humankind" turn is observed in the MOP's broad observer provision. Article 10.4 (Convention on Access to Information, Public Participation in Decision-making and Access to Justice in Environmental Matters, 1998) renders eligible for MOP observer status the UN, its specialised agencies, the IAEA, qualified IOs and non-Party states and economic integration organisations eligible to sign Aarhus. Significantly for civil society, Article 10.5 (Convention on Access to Information, Public Participation in Decision-making and Access to Justice in Environmental Matters, 1998) enables broad observer participation for NGOs: unless at least one third of Parties object, any qualified NGO, expressing intention to participate as an observer, "shall be entitled" to do so. A presumption of nonstate participation is enshrined, with significant objection being necessary for NGOs to be prevented from holding MOP observer status.

Furthermore, and notably for the logic of solidarisation, NGOs "are entitled to seek to address the meetings of the Parties under each agenda item and, having made such a request, are to be included on the list of speakers" (Ebbesson et al., 2014: 215). This indicates a limited but nonetheless gratifying parity of actorhood between state and nonstate actors. This point is reinforced by the archives, which indicate excellent civil society attendance in all MOPs, firmly in line with the precedent set during propagation (UNECE, 2002, 2005a, 2008, 2011a, 2014c, 2017b, 2022b; see this book, Chapter 3). Such a customarily warm disposition towards civil society is not isolated to Aarhus (Ebbesson et al., 2014: 216), but nevertheless serves to reinforce the cosmopolitan impulse, driving human empowerment and co-creation, witnessed throughout the book.

This turn towards a richer, more "humankind" approach in Aarhus' infrastructure is also elicited from the rhetoric and outcomes of the successive MOPs. It is prudent to return to the first MOP in Lucca, Italy, in 2002. The resultant Lucca Declaration (UNECE, 2004c: 1; emphasis added) recognised public engagement as "*vital* for creating an environmentally sustainable future. Governments alone cannot solve the major ecological problems of our time. Only through building partnerships *with* and *within* a *well-informed and empowered civil society*, within the framework of good governance and respect for human rights, can this challenge be met". Lucca (UNECE, 2004c: 5; emphasis added) celebrated "the *constructive spirit* and *close cooperation* among stakeholders which have characterised the processes associated with the Aarhus Convention".

It is instructive, and emblematic of Aarhus' co-creative disposition, that Lucca was accompanied by a parallel civil society conference organised by European ECO Forum, which adopted the Carignano-Lucca Declaration. Carignano-Lucca (European ECO Forum, 2002: 1) hailed "the significant contributions that ECOs and their experts...made to the negotiation, implementation, and ratification of the Aarhus Convention". Lucca and Carignano-Lucca, in both reinforcing the value of co-creation, built further upon Aarhus' propagation and germination, safeguarding the participatory precedent that had been set (see this book, Chapter 3).

Similar rhetoric was evident at the second MOP in Almaty, Kazakhstan in 2005 (UNECE, 2005a). In a speech given there on his behalf, then UN Secretary-General Kofi Annan (UN, 2005: 2; emphasis added) "welcome[d] the active participation of civil society organisations in the Aarhus processes...Your involvement will ensure that the Convention remains a living, evolving instrument. You may not always be happy with the decisions that are taken – that is normal in an open, multistakeholder forum – *but your criticisms and contributions are an essential part of a healthy process*". Aarhus' transition from germination to growth was cast as inseparable from state-human partnership. Unless this "healthy process" ensued, Aarhus would not, to the fullest extent possible, grow and flourish. The Almaty Declaration (UNECE, 2005b: 4; emphasis added) reinforced this rhetoric, encouraging "the public to make full use of its rights under the Convention and recogniz[ing] the role that *all partners* in civil society have to play in its effective implementation. In particular, we welcome the important contribution [of] non-governmental organisations...and call upon Governments and others in a position to do so to give appropriate support, including financial support, to such organisations".

The resultant Almaty Guidelines, on promoting Aarhus' principles in international forums (UNECE, 2005c), sought to operationalise Article 3.7 (Convention on Access to Information, Public Participation in Decision-making and Access to Justice in Environmental Matters, 1998), which obliges Parties to promote Aarhus' principles in other international forums. The Guidelines adopted the same tripartite structure and ethical inclinations as the Convention itself, applying "to all international stages of any relevant decision-making process in matters relating to the environment" (Ebbesson et al., 2014: 69). The Almaty Guidelines are thus prescient for solidarism as they seek to internationally expand Aarhus' concerns for procedural propriety in domains beyond Party jurisdictions. It is also instructive that the Guidelines require such expansion to be sensitive to the risks of global injustice. This is evident in their (UNECE, 2005c: 6; emphasis added) provision that

> Where members of the public have differentiated capacity, resources, socio-cultural circumstances, or economic or political influence, *special measures* should be taken to ensure a *balanced and equitable* process. Processes and mechanisms for international access should be designed to promote transparency, minimize inequality, avoid the exercise of undue economic or political influence, and facilitate the participation of those constituencies that are *most directly affected* and might not have the means for participation without encouragement and support.

An enshrinement of the all affected principle (Ball, 2006; Linklater, 2007; Warren, 2017) arises again, and is reinforced by the Guidelines' (UNECE, 2005c: 7; emphasis added) requirement to include, in such broad international participation, "(a) the members of the public who *are*, or *are likely to be*, *most directly affected*; (b) representatives of public-interest organisations, such as environmental citizens' organisations; and (c) representatives of other interests that might cause, contribute to, be affected by or be in a position to alleviate the problems under discussion". Emphasis is placed on empowering those who lack the privilege that most easily affords influence in decision-making practices. Pursuit of equity and inclusion is bolstered by the Guidelines' (UNECE, 2005c:

6) reinforcement of Aarhus' "nationality, citizenship or domicile" nullification (Convention on Access to Information, Public Participation in Decision-making and Access to Justice in Environmental Matters, 1998: Article 3.9), and their (UNECE, 2005c: 6) encouragement of capacity building "to facilitate international access for the public concerned, in particular NGOs promoting environmental protection, and especially in developing countries and in countries with economies in transition".

Whilst the third MOP in Riga, Latvia in 2008 took stock of Aarhus' achievements in its first decade, and sought longer term implementation strategy, challenges were identified. They related to various needs for implementation enhancement, not least in relation to the adequacy of specified decision-making procedures; the realisation of erstwhile ambitions; the need for transboundary application of the Convention; the need for enhanced quality assurance in public participation, and the removal of impediments to access to justice (UNECE, 2008: 2). There was a sense that Aarhus' full transition from germination to growth would not be possible unless Parties demonstrated more sincere commitment to partnership working, pursuant to the *spirit* as opposed to the mere *letter* of the law.

These concerns continued to be expressed at the fourth MOP in Chisinau, Moldova in 2011. The resultant Chisinau Declaration (UNECE, 2011b: 2; emphasis added) noted, with some degree of urgency, that

> Worldwide, social, economic and environmental challenges are becoming increasingly complex and interrelated. This fact should not discourage the public from involvement in decision-making. Governments must provide the necessary stimulus, tools, information and assistance to enable transparent decision-making processes in order to ensure informed, balanced and effective public participation. Making decisions and decision-making processes fully accountable to the public whom they should serve should become *essential* and not only *procedural*.

This latter point receives italicisation with good reason, for it enjoins states to ensure that, rather than "running through the motions" of procedural compliance, they root public accountability-holding at the very core of their implementation efforts. There was a sense that if Parties positioned their implementation duties as essential primary obligations, as opposed to mere procedural "tick-box" compliance exercises, real gains could be achieved in Aarhus' growth. The Chisinau Declaration's final paragraph (UNECE, 2011b: 3; emphasis added) speaks most acutely to this foregrounding of obligation:

> We consider that our work in implementing the Aarhus Convention is paving the way for a *universal application of Principle 10*. While recognising that there are different ways to implement the principle, we offer to share our experience with all countries that wish to join the Aarhus family, to replicate its achievements or to be inspired by this most ambitious venture in environmental democracy undertaken under the auspices of the United Nations.

Bold globalising rhetoric is evident, and it speaks to the possibility of using the Convention as a legal springboard from which to attempt something radically ambitious: the development of a "global Principle 10" instrument. The rhetoric was reinforced by Chisinau's high level sessions, on advancing sustainable development in the UNECE region, and on the global promotion of Principle 10. At this point, Aarhus' transition from germination to growth had been consolidated. Instructive evidence, in this regard, is found in a speech given by a chief Aarhus stakeholder (Wates, 2011a: 2) to Chisinau's high level session on Principle 10, which noted that "there is one thing that this [UNECE] region can offer the rest of the world, something which does not cost the Earth. It can offer the Aarhus Convention. It can offer an important tool for a more participatory, transparent form of governance". Aarhus was portrayed as the vehicle with which to codify Principle 10's globalisation. Wates (2011a: 3) enjoined Aarhus' state Parties to

> affirm their support for the development of regional conventions on environmental democracy in other regions of the world, following the example of Aarhus. The willingness of UNECE's sister organisation, the Economic Commission for Latin America and the Caribbean (ECLAC), to broker the development of a legally binding instrument in the Latin American region is very much to be welcomed in this context... The Aarhus Parties should also reiterate in the Rio+20 process their willingness to support any initiative to develop a global instrument, again offering the Aarhus Convention as a useful model.

Chisinau's ambition to pollinate Aarhus overseas was borne in reality. On top of UNEP's (2015) Bali Guidelines for the Development of National

Legislation on Access to Information, Public Participation and Access to Justice in Environmental Matters, the Escazú Agreement (Regional Agreement on Access to Information, Public Participation and Justice in Environmental Matters in Latin America and the Caribbean 2018; see Etemire, 2023a, 2023b) contributed to a relative replication of Aarhus' provisions in the ECLAC region. Of further significance is Guinea-Bissau's accession to Aarhus (United Nations, 2023): Guinea-Bissau was the first, and to date remains the only, state outside the pan-European region to have acceded to the Convention. Whilst Aarhus' strategic tier previously paid attention to potential accessions by Mongolia and Morocco (UNECE, 2014c: 11), neither had acceded at the time of writing (United Nations Treaty Collection, 2023).

This drive towards growth had only intensified by the time of the fifth MOP in Maastricht, The Netherlands in 2014. There, the UN Secretary-General (UN, 2014: 1) praised Aarhus for having "inspired similar initiatives that intend to replicate their achievements beyond the European region" and for having provided "valuable lessons on how governments can effectively engage different stakeholders in decision-making…to contribute to our broader campaign for a sustainable future". The subsequent Maastricht Declaration (UNECE, 2014a: 2) reinforced a "strong commitment to empowering people with the right to a healthy and favourable environment, improving the state of the environment and to furthering sustainable development through the promotion of the Convention and its Protocol and their principles within the ECE region and beyond". Whilst it remains a declaration, it nonetheless plays a valuable role rhetorically, at least, in *signalling* substantive environmental rights and maintaining the possibility of further replication of the Convention.

It is also quite telling that renewed attention was paid, at Maastricht (UNECE, 2014a: 3), to global injustice, not least in recognition that Aarhus' promotion in international forums remained "essential…not only for forums dealing directly with environmental matters…but also for related forums, such as the international financial institutions and trade-related organisations, where transparency ... is of the utmost importance". Such renewed attention should be understood in the context, at that time, of the development of the UN's SDGs, adopted by all UN member states in 2015 to set global goals for achieving sustainable development (United Nations Department of Economic & Social Affairs, 2023). The SDGs built on the arguably less successful MDGs

(Briant Carant, 2017; see also Death & Gabay, 2015). Application of Aarhus' provisions in international forums, not least those relevant to discrete endeavours towards sustainable development, would contribute, I think, at least to the fourth (Quality Education), sixteenth (Peace, Justice and Strong Institutions) and seventeenth (Partnerships) SDGs (UNECE, 2017a: 3–4; 2022a: 4). To deem Aarhus, as I do, a valuable tool for ensuring that the SDGs are less chimeric than the MDGs is not too controversial (Atapattu, 2019; Orellana, 2016; Zhao & Butcher, 2022). After all, it was Maastricht itself (UNECE, 2014a: 4–5; emphasis added) that found, in Aarhus' internationalisation, potential for "setting a *people-centred* post-2015 development agenda".

However, the archives indicate that not all was well, in the strategic tier, more recently. A risk of statist, to a degree even realist, regression is evident. Universal applause of the state of implementation eluded the sixth MOP in Budva, Montenegro, in 2017. European ECO Forum expressed, there, "concern about the speeding up of the public participation process" in some Party jurisdictions, plus "the need to ensure the participation of marginalised groups" (UNECE, 2017b: 6). European ECO Forum (UNECE, 2017b: 18) also regretted Parties' failure to "systematically ensure...that their representatives in international forums understood the implications of the Convention and duly upheld their obligations to promote its principles when negotiating in the context of relevant forums".

Moreover, and foretelling Belarus' eventual withdrawal from the Convention (see this chapter, below), the EU expressed concern about the plight of environmental civil society actors in Belarus (UNECE, 2017b: 22). Foreboding and worry are apparent in the Budva Declaration (UNECE, 2017a: 2; emphasis added), which noted "with great concern that human rights violations, terrorism and radicalism are undermining the foundations of democracy in many countries ... *including* in countries that are party to the Aarhus Convention and its Protocol and other ECE countries". With a tone of solemnity and urgency, Budva (UNECE, 2017a: 2; emphasis added) was "*particularly alarmed* by the increase in the harassment, silencing and *even murder* of environmental activists around the world", calling on Parties "to ensure due protection of environmental activists, whistleblowers and NGOs so that they can exercise their rights under the Aarhus Convention and its Protocol...*without being threatened in any way*". This is a far cry from the bold pioneer rhetoric of celebration and growth that accompanied the first decade following

Aarhus' entrance into force. It indicates major concerns surrounding the mortal risks facing what are now termed environmental defenders.

Risks of regression continued to surface as the twentieth anniversary of Aarhus' entrance into force was marked at the seventh MOP in Geneva, Switzerland, in 2021. Nothing is more insightful, than the need for the MOP to have adopted Aarhus' rapid response mechanism for the protection of environmental defenders. I refer to the *need* to have adopted the mechanism, because there was no anticipation, during propagation, that such a need would arise. Article 3.8's prevention of "persecution, penalisation and harassment" (Convention on Access to Information, Public Participation in Decision-making and Access to Justice in Environmental Matters, 1998) was deemed enough. There was never any indication that the state of implementation, and the risks facing humans and their associations, would be such that rapid response interventions would be necessary. Whilst the rapid response mechanism is a welcome safeguard for those exercising their Aarhus rights, it cannot be denied that it constitutes a reaction to the plight of persons exercising their rights *inside* Party jurisdictions. In ideal circumstances, this mechanism would not be required, because environmental defenders would do their work without impediment or mortal danger. The realities, however, are different. It is a point of regret that such necessity arose. Over a decade ago, I studied the demise of one "Aarhus activist" (Weaver, 2015: 190–194); since then, it appears that the lot of environmental defenders scarcely improved, thus necessitating the mechanism (Ryall, 2023; Special Rapporteur on Environmental Defenders, 2022; UNECE, 2022b: 14; Weber, 2022).

With this in mind, the rapid response mechanism seeks to afford immediate relief when defenders are "harassed, prosecuted or penalised" (Weber, 2022: 75). A Special Rapporteur on Environmental Defenders operationalises it; their task "is to take measures to protect environmental defenders who suffer harmful acts or are under the imminent threat of such acts" (Weber, 2022: 75). Humans are empowered to the extent that they can directly lodge concerns with the Special Rapporteur, either for themselves or on others' behalf; Parties and the Secretariat can also raise concerns. Unlike the compliance mechanism's proceedings (see this chapter, below), domestic remedies do not need to be exhausted before a communication is made to the mechanism (Weber, 2022: 75). The Special Rapporteur can place immediate or ongoing protection measures on Parties, in order for them to "take action, or…refrain from action, in order to protect named persons from persecution, penalisation or harassment" (Ryall, 2023: 163). And the Special Rapporteur can take measures that might relate to the employment of diplomacy, the dissemination of public statements and the cooperation with cognate human rights organs (Ryall, 2023: 162). Around a dozen complaints were received by the Special Rapporteur at the time of writing.

It is in this context—of the risks facing environmental defenders, and the seventh MOP seemingly having no option other than to initiate safeguards—that the cracks leading to Belarus' eventual withdrawal from the Convention began to show. When adopting the rapid response mechanism, the seventh MOP agreed to hold an extraordinary session in 2022 to elect an independent Special Rapporteur on Environmental Defenders. Bearing in mind Aarhus Convention Compliance Committee case ACCC/C/2014/102 (Aarhus Convention Compliance Committee, 2017), concerning alleged persecution of civil society activists by Belarusian authorities, it is instructive that a Belarusian delegate stated, at the seventh MOP (UNECE, 2022b: 14), that "Belarus reserves its position that it may not recognize the candidature of a special rapporteur not elected through consensus". The dogged voluntarism of sovereignty resurfaced. Belarus "gave notice", in effect, that it would not recognise any such Special Rapporteur's authority. Belarus demonstrated its ability and willingness to withdraw from one element of the diplomatic "pitch", having perceived the Convention to have surpassed its threshold of tolerance. But in a rejoinder invoking the spirit of the law, Belarus was rebutted not by a state but by a representative of civil society (UNECE, 2022b: 14): a delegate of European ECO Forum noted "that the Convention's rules of procedure applied to all Parties, and that was fundamental to the international law of treaties, ensuring the integrity of process". No more poignant an indication is found of the tension between the prerogatives iterated by a state disquieted by the intrusiveness of environmental multilateralism, and a civil society delegate whose voice was recorded with parity.

Belarus' ultimate withdrawal from the Convention entered into force in autumn 2022, something which the European ECO Forum "deplore[d]" and about which it expressed "deepest concern" (European Environmental Bureau, 2022). What coincided with this unilateral withdrawal was not only Belarus' implication in the Russian Federation's illegal invasion of Ukraine, but the liquidation of Belarus' oldest environmental civil society organisation *Ecohome*, amongst many other such organisations, unions and media establishments (see European Parliament, 2022).

The risks of regression should be clear. The outcome of the seventh MOP, the Geneva Declaration (UNECE, 2022a: 2), warned of "serious challenges", of "social conflict and political unrest that can lead to people living under the threat of harassment or even in fear of their life". Geneva (UNECE, 2022a: 6; emphasis added) was "*alarmed* by the serious situation regarding the persecution and harassment of environmental defenders in the ECE region and beyond…We acknowledge that the *chilling* effect this creates *discourages* public participation and *prohibits* access to environmental justice, *thwarting* the aims of the Convention". This disquieting and dark rhetoric was augmented by Geneva's (UNECE, 2022a: 5; emphasis added) recognition that a "majority" of Parties were engaged with China's Belt and Road Initiative (BRI), and that appropriate "*safeguards*" for maintaining Convention compliance would be "*critical*" in this regard. It is unfortunately the case that a return of "hard" geopolitics, in recent years, appears to have necessitated such stern rhetoric, in the strategic tier of the infrastructure.

To conclude, the strategic tier is speckled with glimpses of cosmopolitanism, but remains consistent with the orthodoxies of multilateralism. A certain "humankindness" is evident in the MOP's customary warmth to civil society, and in its frequent invocations of co-creation. Solidarism was evident in the internationalising (but not globalising) attempts to pollinate Aarhus overseas, as well as in the concerns expressed for global injustice: for amplifying human voice in pursuit of sustainable development.

Yet tension exists between the solidarist ambitions and pluralist constraints of the strategic tier. The persistence of state consent is thrown in stark relief by the NIR challenges. The doggedness of sovereignty was evident in the tensions between the letter of the law, emblematic of pluralist minimalism and the more solidarist spirit of the law. From the above, it is clear to me that states will embrace the spirit of the law, and *permit* its embrace, only to the extent that it serves their purposes. Solidarist ambition will only be allowed to grow to the extent that it remains firmly in the confines of the consenting states' thresholds of tolerance.

Certain developments, in this regard, are troublesome, and they raise my concerns about the risks of regression. Concerns relating to the plight of environmental defenders arguably left the MOP with no option other than to initiate a rapid response mechanism. Belarus' unilateralism is a bruise on the Convention, and a reminder that state consent has gone nowhere. Similarly, Parties' geopolitical and economic bargaining, for instance with the Belt and Road Initiative (see Flint & Zhu, 2019; Nye, 2023: 109–111; Yu, 2019) resulted in them having, in effect, to be warned about their involvement and reminded of their environmental and democratic responsibilities. I sense a darkling tone in the more recent archives, and it only serves to reinforce the risks that come with the return of hard geopolitics in Europe.

The implications for solidarisation relate to the mutual tensions between state and human prerogatives. Whilst the rhetoric, in the strategic tier, was often animated by a cosmopolitan impulse of "humankindness", ethical progress was tempered by the predominance of sovereignty. The heavy counterbalancing in favour of sovereignty is significant and unlikely to lighten in the strategic tier. States continue to prevail in the "high political" arm, flying the flags of their sovereign prerogatives in the negotiating chambers of an inter*national* environmental agreement.

Operational Tier: Working Group of the Parties, Bureau, Task Forces and Secretariat

It will come as little surprise that Aarhus' operational tier is also consistent with state-oriented multilateralism. I regard the operational tier—the Working Group of the Parties, Bureau, Task Forces and Secretariat—as comprising Aarhus' vital organs, sustaining its functional continuity between MOPs. The Working Group of the Parties was established by the first MOP under Decision I/14 (Banner, 2015: 5). It oversees the implementation of Aarhus' intersessional work between MOPs; conducts preparatory work for MOPs; monitors the need for amending the Convention; makes recommendations to the MOP for purposes of achieving Aarhus' objectives and discharges any duties requested by the MOP (Banner, 2015: 5). The Working Group of the Parties is thus an instrument with which to sustain and serve the Convention. To substantiate this, one need only reflect on the Working Group's most recent (at the time of writing) twenty-seventh meeting, held in June 2023 (UNECE, 2023d). There, the implementation and promotion of the trinity's provisions was reviewed, as was the work towards ratifying the GMO amendment. Financing was reviewed, as well as the international promotion of Aarhus. The meeting assessed the implications of the Russian invasion of Ukraine for Ukraine's implementation of the Convention

(UNECE, 2023d). The customary warmth shown towards civil society had gone nowhere; ample opportunities were provided for nonstate actors to voice themselves, their opinions and their expertise. What ensued was a near-three-day dialogue between state and nonstate actors, on the state of implementation and the road forward. The sustained actorhood enjoyed by environmental civil society was demonstrated in the fact that the European ECO Forum was responsible for coordinating the interventions of the various nonstate actors present (UNECE, 2023d).

Similarly warm dispositions are also evident in the Bureau. A critical decision-making body, it meets more often than the MOP. It is designed to be agile in responding to operational needs arising during implementation. Its members are elected by the MOP; significantly the Bureau is mandated to "invite a representative of non-governmental organisations established for the purpose of, and actively engaged in, promoting environmental protection and sustainable development…to attend bureau meetings as an observer" (UNECE, 2023b).

Meanwhile, Aarhus' task forces are integral to the development of Aarhus' practical implementation. Three function at the time of writing: one for each procedural pillar (see this book, Chapter 4). Together, the task forces facilitate exchange of information, best practice and capacity building; analyse technical developments and pertinent themes arising in their respective pillars' implementation; address common issues and contribute to the attempted enhancement of implementation. It has been observed that the task forces collectively constitute "a forum for the free exchange of ideas" and their arrangements have been "praised as facilitating discussion, and particularly for giving *greater room for nonstate actors to speak*" (UNECE, 2013: 28; emphasis added). The task forces have in fact been found to constitute "the main mechanism for progress during intersessional periods" and "facilitators for horizontal reporting" (UNECE, 2013: 32) between concerned stakeholders. I thus regard the task forces as dialogue forums in which state and nonstate actors can engage and co-create on a relatively equal footing, in order to sustain the Convention's growth. I should also remark that whilst environmental civil society and law tend to be well represented in the operational tier, it is also the case that industry representatives avail themselves of the opportunity to observe, and participate in, the Convention's working life as well.

None of this machinery would function as smoothly as it does without the secretariat. The latter is much aligned with its counterparts in other international environmental agreements. The secretariat consists of international civil servants whose role is to practically and logistically support Aarhus' daily life and work. It prepares and administers relevant documents; assists Parties and other actors; receives and disseminates reports and maintains communications with stakeholders in order to preserve as smooth a coordination and administration of the agreement as possible (see Bauer et al., 2012: 34; Churchill & Ulfstein, 2000: 627). As we shall see in the next section, Aarhus' secretariat is also entitled to refer non-compliance matters to the Compliance Committee. Formally and theoretically, then, the secretariat could trigger compliance proceedings. However, as I complete this book, no such secretariat-initiated referrals were made (UNECE, 2023c).

Having briefly treated the operational tier, my following thoughts arise. It comes neither as disappointment nor surprise that the operational tier is state-oriented. Little further can be done to achieve radical progress within the well-worn machineries of multilateralism. Other than maintaining its customary dialogic warmth, I struggle to identify more robust modalities of "humankindness" that can be developed within the confines of UN-administered international law and diplomacy. Civil society benefits from a relative parity of actorhood, its voice coordinated and articulated. But Aarhus' ambitious intersessional work nonetheless remains administered by a "tried-and-tested" machinery that relies on the blessing of consenting sovereign states.

The Tactical Tier: Aarhus' Compliance Mechanism

Whilst glimpses of cosmopolitanism were observed in the strategic and operational tiers, both are heavily state-oriented, and understandably so. The "high politics" and operational "life support" of international environmental agreements are inherently statist, requiring as they do sustained collective investments of sovereignty, by states, for their continuity. It is understandable not only that states prevail, but that they will *expect* to prevail, over environmental multilateralism. The real solidarist magic, though, is achieved in Aarhus' compliance mechanism. I regard the mechanism as tactical because its role is to establish, in highly specific cases, whether a Party acted in compliance with the Convention or not. Before proceeding, let me clarify that compliance mechanisms are built into international environmental agreements in order to support Parties experiencing compliance difficulty, and to address highly specific issues of non-compliance (Koester, 2015: 201). This latter term is popularly employed, as opposed to “contravention”, “violation” or “infringement”, because “non-compliance” adopts a relatively more optimistic assumption that Parties “basically wish to comply with their obligations” and that “non-compliance is a result of a lack of ability or capacity…rather than of a deliberate decision to ignore obligations” (Koester, 2015: 201).

The very rhetoric of compliance, then, is premised on an understanding of international environmental politics and law that presupposes the value of supporting states that wish to perform their duties more effectively. Presciently for our study of solidarisation, compliance mechanisms have been noted for their value as supplements to the “age-old attempts of international law to create free, yet disciplined and responsible sovereigns” (Aalberts & Werner, 2011: 2196). As such, it should come as scarce surprise that such mechanisms were noted for their solidarist quality. Falkner (2021: 207), for instance, observed “through a solidarist lens” compliance mechanisms’ proclivity to “help states meet their treaty obligations by strengthening domestic capacity for implementation and information exchange”.

Aarhus’ compliance mechanism was established under Article 15 (Convention on Access to Information, Public Participation in Decision-making and Access to Justice in Environmental Matters, 1998) and mandated by Decision I/7 at the first MOP (UNECE, 2002). Under Article 15, the compliance mechanism is to be “non-confrontational, non-judicial and consultative” (Convention on Access to Information, Public Participation in Decision-making and Access to Justice in Environmental Matters, 1998). Noting earlier coverage (see this chapter, above) of the discords between the “letter” and more expansive “spirit” of the law, it is worth observing here such tensions between the non-judicial legal wording and the reality of compliance under the Convention (Fasoli & McGlone, 2018). This will be scrutinised later. But for now, the non-judicial wording was, on my understanding, a “lowest common denominator” outcome of Aarhus’ negotiations. In other words, the meek non-judicial terminology constituted a concession to certain states engaged in bargaining before the final draft was agreed. Indeed, these non-judicial arrangements were regarded by the UK, at the time, as “the weakest text on non-compliance in any international environmental law” (in Brady, 1998: 73). Such criticism was not flattering in any respect. However, as we shall see, Aarhus’ compliance mechanism is actually *not* as non-judicial as we might infer.

Shortly, I will elicit three solidarising indicators from the mechanism—the Compliance Committee's composition, its quasi-judicial practices and the provision for humans to directly raise non-compliance communications. But beforehand, a useful vignette from the first MOP (UNECE, 2002) is due. There, the US—not Party to Aarhus—issued a statement on the compliance mechanism's creation. Adopting a tone of unilateral reticence, the statement came five months after a similar declaration of the US' intention not to become Party to the Rome Statute of the ICC, and of its having "no legal obligations arising from its signature on December 31, 2000" (United States Department of State, 2002; see Coalition for the ICC, 2023). The US (UNECE, 2002: 19) observed "curious and troublesome elements of the compliance regime", stating that it did "not consider the compliance rules adopted…to be a precedent for compliance procedures in other regional or multilateral environmental agreements". Such reticence was attributed to "a variety of *unusual* procedural roles that may be performed by nonstate, non-Party actors, including the nomination of members of the Committee and the ability to trigger certain communication requirements by Parties under these provisions" (UNECE, 2002: 19; emphasis added).

This vignette is instructive because it demonstrates the persistence of state consent: the persistence of states' ability and willingness to enter and leave the diplomatic "pitch" at will. This chimes with the argument sustained throughout this book: that states will permit ambitious solidarist progress only to the extent that it serves their interests. It is the first prerogative of sovereignty that states will only "invest" it in the multilateral arena to the degree that they will achieve a "return on investment", in terms of benefits accrued to their national interests, territories, populations, economies, security priorities, etc. In line with the logic of solidarisation, I have argued throughout that solidarist ambition is necessarily delimited by sovereignty, and thus kept "in check" within the threshold of states' tolerance. As such, the US' unilateralism is not wholly surprising. It is not the first instance of unilateralism raised in the book: we should recall the concurrent eligibility and refusal of Canada and the US to negotiate and sign Aarhus (see this book, Chapter 3); the Russian Federation's negotiation of, and influence over, the nascent Convention before its "last minute" avoidance of signature (see this book, Chapter 3) and in Belarus' withdrawal (see this chapter, above).

Yet the US vignette that I have just employed also indicates something radical about the compliance mechanism: *so* radical that the US deemed it necessary to express such reservation in the first place. The three solidarising indicators that follow will shed light on such unilateral reticence, and on why the Convention was recently cast as an innovator in "private environmental enforcement" that enables the functioning of environmental citizen "watchdogs" (Kingston et al., 2022: 2).

Composition

The first indicator of solidarism is the Compliance Committee's employment of autonomous human individuals who are not bound by governmental allegiance. They are recruited for their expertise, not their state-oriented power and privilege. Their findings and contributions, regarding compliance, should thus be detached from undue political influence. As such, whilst these nine Compliance Committee members must be nationals of Parties or Signatories to the Convention, they serve in a personal capacity (UNECE, 2019: 18), thus helping to preserve the compliance mechanism's objectivity. This is important, because Compliance Committee findings yield influence on domestic change inside Parties' jurisdictions.

Another safeguard, in this respect, is the possibility for Compliance Committee candidates to be nominated not only by Parties and Signatories, but also by NGOs. This was deemed "unprecedented" (Morgera, 2005: 140), early in Aarhus' entrance into force. Nonetheless, and in line with the sovereign prerogative, states retain a final grip on the Compliance Committee's composition, as members are elected consensually by the MOP, or by secret ballot if consensus is not reached (UNECE, 2019: 18). An ultimate check is therefore placed on the Compliance Committee's composition by states. After all, it is in states' interests to retain sovereignty *over* a mechanism that is designed to hold Parties accountable for their compliance.

Furthermore, an attempt is made to safeguard diversity and representativeness vis-à-vis the Compliance Committee's composition. We may consider, in this regard, the provision that the Compliance Committee may not include more than one national of the same state, and that the geographical distribution of Compliance Committee membership, as well as diversity of experience, need to be accommodated accordingly (UNECE, 2019: 19). The focus is clearly on establishing a Compliance Committee of experts who serve as such: experts who are not chained to governmental allegiances, and who are reflective of the broad "Aarhus community" in discharging their responsibilities of establishing whether or not a Party acted in compliance with the Convention, in a given set of circumstances.

Quasi-Judicial Working Practices

The second solidarist indicator speaks to a discord that I raised at the start of this section, between (a) the "non-confrontational, non-judicial and consultative" (Convention on Access to Information, Public Participation in Decision-making and Access to Justice in Environmental Matters, 1998: Article 15) letter of the law, and (b) the more judicial realities of the compliance regime. To express this discord further, let us follow Wates' (2011b: 394) position that Aarhus' compliance mechanism is not, and nor is it intended to be, "a redress procedure with the powers to undo all wrongs committed against the ideals of environmental democracy". Neither does the mechanism have the competence to "overrule decisions of national courts or administrative authorities, abrogate national laws...intervene directly on a communicant's behalf" or compel "damages to be paid as a result of a Party's non-compliance" (Koester, 2015: 204). It would be inaccurate to deem the compliance mechanism a court per se.

But I would find it difficult to deny that the mechanism is *quasi*-judicial. Aarhus' compliance process shares some close affinities with a court trial, given its codification of processual propriety, the use of what can only really be described as a lengthy multistage hearing (if not a trial), and the meticulous, evidenced formulation of compliance findings that contribute to a body of "case law" which, as we saw earlier (see this book, Chapter 3), is cited by international courts. Whilst Compliance Committee findings are theoretically at risk of political influence, given that the MOP is responsible for endorsing them, they nonetheless remain founded upon conclusions of a "strictly legal character" (Koester, 2015: 203). Such quasi-judicial quality is further substantiated by Day's (2015: 188) finding that whilst communicants of alleged non-compliance are under no obligation to use legal representation, "some Parties routinely instruct counsel", thereby intensifying "the pressure on Communicants to do likewise in order to avoid an inequality of arms". The rhetoric here is as adversarial as that of court litigation. At the very least, it is significantly detached from the consultative tone implied by the letter of the law.

We can further substantiate the mechanism's quasi-judicial character with recourse to the procedural precision of its compliance assessment process (Fasoli & McGlone, 2018; Koester, 2005, 2007, 2015; Kravchenko, 2007; Samvel, 2020; UNECE, 2019, 2022c). Even before a communication is assessed, the Compliance Committee considers whether domestic legal remedies were used by the communicant. There is, then, a concern at the very outset for establishing whether domestic due process was followed inside Party jurisdictions. If a case is admitted, a hearing ensues between the communicant and the Party whose compliance is under scrutiny. This entails the delivery of presentations by both sides; the articulation of Compliance Committee questions and responses; the expression of observers' comments and the delivery of closing remarks by both sides. This modality of due process involves impartially gauging the facts of the matter, in order to verify or nullify alleged non-compliance. Findings and recommendations are drafted by the Compliance Committee, concluding as to whether or not a Party acted in compliance, in the given circumstances. Responses are invited before drafts are reviewed, edited and finalised. The Compliance Committee will then adopt its findings, which are considered by the MOP. If a Party is found to have been in a state of non-compliance, and if the MOP agrees, the MOP will endorse the findings and make the recommendations suggested by the Compliance Committee. This is important, because under international law (Vienna Convention on the Law of Treaties, 1969: Article 31.3.a), such endorsement may constitute Parties' agreement on the interpretation and application of the international agreement, which would need accounting for in future proceedings. It is in this context that the mechanism has generated a wealth of "binding" "case law" (Andrusevych et al., 2011; Fasoli & McGlone, 2018: 45; Jendrośka, 2012: 75; Krämer, 2012: 98; Samvel, 2020: 233–234). It is instructive that the most recent seventh MOP endorsed each of the Compliance Committee's findings of non-compliance (Ryall, 2023: 164).

Pursuing this quasi-judicial theme further, it is the MOP's duty to "decide upon appropriate measures to bring about full compliance", and it can adopt one or more of the below options, on considering the Compliance Committee's findings (UNECE, 2004a: 7–8):

a. Provide advice and facilitate assistance to Parties
b. Make recommendations to Parties
c. Request Parties to submit compliance strategies

d. Make recommendations to Parties on measures to address issues raised by the public
e. Issue declarations of non-compliance
f. Issue cautions
g. Suspend a Party's rights and privileges
h. Take other non-confrontational, non-judicial and consultative measures as appropriate.

However, we would do well to note that—*with a Party's consent*—the Compliance Committee can autonomously initiate compliance activities "[p]ending consideration by the Meeting of the Parties, with a view to addressing compliance issues without delay" (UNECE, 2004a: 7). In an Annex to Decision I/7 (UNECE, 2004a: 7; emphasis added), the Compliance Committee may, "*in consultation*" with a Party, "provide advice and facilitate assistance" on Parties' compliance. And it may, "*subject to agreement*" with a Party, "make recommendations to the Party concerned; Request the Party concerned to submit a [compliance] strategy", and, in cases where the public lodges communications, "make recommendations" to the Party "on specific measures to address the matter raised" (UNECE, 2004a: 7; emphasis added).

As such, ultimate "ownership" of compliance enforcement rests with the MOP. Whilst the Compliance Committee can instigate compliance activity, it can do so only with state consent, and only using limited options. I regard this as an instance of counterbalancing in favour of sovereignty, a check upon the "runaway" unshackling of the compliance mechanism from its sovereign masters. The above findings also need caveating with the fact that Aarhus' ambitious compliance provisions are not alone in international law (Fasoli & McGlone, 2018: 36).

Yet it would be a disservice to downplay the mechanism's quasi-judicial qualities too much. We should not deny the existence of a clear tension between the pluralist letter of the law—providing for Aarhus' non-confrontational, non-judicial and consultative compliance mechanism—and the above court-like concerns for due process, procedural precision and factfinding, with two adversarial parties likely employing legal assistance. There is also a tension between the provision for a non-judicial mechanism and the latter's generation of "case law". Fasoli and McGlone (2018: 33) are spot on when they argue that such "consultative" arrangements, in the letter of the law, should surely "militate against arrangements with legally binding results".

As such, we witness yet another counterbalancing, in this book, between solidarist ambition and the pluralist preservation of sovereign prerogatives. By codifying non-judicial compliance arrangements, the letter of the law seeks to avoid unduly constraining and encroaching upon Aarhus' consenting Parties. But this statist letter of the law exists in tension with the mechanism's more solidarist and intrusive practices, which bruise the minimalism with which states can limit themselves to the letter of the law.

Human Empowerment

The compliance mechanism's third solidarist indicator is rooted in its capacity to directly empower humans and their associations. Whilst Aarhus' compliance proceedings are customarily transparent (Day, 2015: 188; Dellinger, 2012: 324; Koester, 2007: 85; Kravchenko, 2007: 5), we should rather focus on Aarhus' empowerment of humans, NGOs and combinations thereof to directly lodge communications of non-compliance with the Compliance Committee. This has been coined as "highly unusual" (Koester, 2015: 203). Whilst there are five means by which to trigger compliance review—"Party-on-Party" referral, Party self-referral, secretariat referral, MOP request and public communication (UNECE, 2019: 25)—it is the latter "public trigger" that instigated review in the "vast majority" of cases (UNECE, 2019: 25). Indeed, at the time of writing, over 96% of compliance referrals arose from public communications.[1]

This latter "public trigger" is regarded here, as with Article 9.3 (Convention on Access to Information, Public Participation in Decision-making and Access to Justice in Environmental Matters, 1998; see this book, Chapter 4), as being implicated in a broader push towards environmental *actio popularis*, underway in the Convention. Certainly a "novelty" (Peeters & Nobrega, 2014: 362), and in line with the "citizenship, nationality or domicile" nullification (Convention on Access to Information, Public Participation in Decision-making and Access to Justice in Environmental Matters, 1998), the public trigger does not require communicants to be citizens of the Parties about which they lodge communications. Moreover, and as above, communicants may comprise human individuals or groups thereof, NGOs, and also combinations of NGOs and human individuals (UNECE, 2022c). Likewise, NGO communicants need not be based in the Parties about which they have lodged communications. This is deeply empowering, and serves to 'transnationalise' human environmental voice and the associated capacity to initiate proceedings that can yield influence inside Parties' domestic jurisdictions.

Before closing the chapter, let me pass some thoughts on the tactical tier. The compliance mechanism is deserving of continued scholarly and societal attention, as it helps to amplify human voice in an attempt to hold states accountable for their environmental actions. We witnessed effort, in this section, to ensure that the mechanism was diverse and reflective of the broad "Aarhus community". The provision for NGOs to nominate Compliance Committee candidates offered a modest check on the predominance of statist "high politics" over, and a safeguard for including civil society in, Aarhus' compliance mechanism. The latter's quasi-judicial practices were seen to be in tension with the pluralist, more cautious, provision for Aarhus' "non-judicia" compliance mechanism. They were portrayed as court-like and quasi-jurisprudential. But most significant is the provision for humans and NGOs to trigger compliance review. It is here that the real solidarist "magic" occurs. With over 96% of compliance referrals emerging from public communications, human voice takes centre stage in the compliance mechanism. *State* Parties are held accountable by, and before, *nonstate* actors, and must comply with any enforcement requirements that result from such civil society intervention.

Yet the above account is replete with sovereign counterbalancing. We may consider, to this end, the fact that the non-judicial letter of the law was itself a "lowest common denominator" of Aarhus' propagation: a deliberate tempering of the cosmopolitan impulse by consenting states. We might turn to the US' unilateralism vis-à-vis the compliance mechanism: a reticence similar to that expressed by the US when refusing to sign the legal arrangements for the ICC. Counterbalancing is also evident in the MOP's sovereignty *over* the Compliance Committee's composition, and in the need for the MOP to endorse Compliance Committee findings, ultimately rendering compliance susceptible to the "high politics" of environmental multilateralism.

Finally, state consent resurfaces in the need for Parties to agree to autonomous interventions by the Compliance Committee, with an added safeguard against usurpations of sovereignty being found in the need for domestic redress mechanisms to be accounted for during initial compliance assessments. It is in this context that I regard Aarhus as being implicated in a broad push towards environmental *actio popularis*.

However, and firmly in line with the logic of solidarisation that permeated this book, such a significant broadening of human legal actorhood is anchored well within the bounds of sovereignty, and remains contingent upon the willing consent of states.

Conclusion

This chapter assessed Aarhus' organisational infrastructure, partitioning it into strategic, operational and tactical tiers. In many respects, the picture painted was one of mainline multilateralism. Whilst it is tempting to seek a mythical turn towards post-sovereign cosmopolitan governance, the reality is different. Aarhus' organisational infrastructure remains largely state-oriented and continues to hinge upon the voluntarism of state consent. It is an international machinery in which consenting states give their "blessing" to arrangements that travel on well-worn paths of multilateralism.

It is of course the case that glimpses of cosmopolitanism were evident. Where they *were* evident, they reinforced my finding that Aarhus plays a pioneer role in codifying modest human-oriented solidarism. All three tiers demonstrated the customary "humankindness", the inclusive disposition towards partnership and co-creation, that is a signature of the Convention. An invocation of this cosmopolitan impulse was witnessed not least in a relative parity of actorhood enjoyed by nonstate actors, and also in the fostering of human empowerment and amplification of human voice within the infrastructure.

More robust solidarist potential was found in the compliance mechanism's composition, in its capacity to bestow legal "teeth" on the Convention through its quasi-judicial practices, and in its empowerment of humans and NGOs to directly submit communications to the Compliance Committee. I made the case, furthermore, that whilst the mechanism is formally non-judicial, such pluralist minimalism is agitated by the mechanism's "harder", more jurisprudential realities.

Yet sovereign prerogatives continue to firmly reinforce Aarhus' organisational infrastructure. Counterbalancing in favour of sovereignty animated the above assessment. Laggardness was seen to hinder Aarhus' NIR performance; less than satisfactory domestic implementation regimes were seen to have necessitated a rapid response mechanism. The doggedness of sovereignty was witnessed in both Belarus' withdrawal from the Convention, and in the US' refusal to heed the legitimacy of the compliance mechanism. What is evident is a mutual regulation between pluralist statism and solidarist ethical ambition. On one hand, Aarhus' *rationales* are driven by a cosmopolitan impulse of "humankindness": of empowerment, voice and co-creation. This impulse resonates with Gilmore's (2023: 96) idea of a solidarist "co-production" that sincerely values the diversity of human actors and their "day-to-day lived experiences and subjectivities" whilst remaining "critically attuned to pluralism".

Yet on the other hand, Aarhus' *realities* temper the human-oriented solidarism sought. A number of checks against the undermining of sovereignty were observed, serving to signal the persistence of sovereignty in environmental multilateralism. The solidarist progress that *was* achieved, ensued within existing power frameworks, and at a pace and scale deemed acceptable by the consenting states. Sovereignty was reaffirmed as critical to multilateralism; none of the above "humankindness" would occur without state consent.

We saw that states can and do leave the diplomatic "pitch" when they deem solidarist ambitions to have crept too far beyond their thresholds of tolerance. Whilst sovereignty underwent ethical maturation in the above organisational infrastructures, it also served to set the parameters in which Aarhus' rationales are confined. This reinforces the persistence of, and insistence upon, sovereignty in these international society arrangements, underscoring the fact that "even when genuinely consensual, the promotion of solidarist values both *depends on*, and *reinforces*, the power and privileges of the dominant state or states" (Hurrell, 2007: 305). To this extent, my findings broadly align with Ahrens' (2019: 287) observation that in an international society residually characterised by "considerable pluralist ideas, principles and structures, any solidarist agenda cannot but account for pluralist reservations".

Note

1. At the time of writing, UNECE (2023c) lists 208 compliance cases: 2 referrals by Parties about Parties (0.9615%), 1 Party's self-referral (0.4808%), 0 secretariat referrals (0%), 5 requests by the MOP (2.4038%) and 200 public communications (96.1538%). The overwhelming majority of Aarhus' compliance cases thus arise from public communications.

Bibliography

Aalberts, T., & Werner, W. (2011). Mobilising uncertainty and the making of responsible sovereigns. *Review of International Studies, 37*(5), 2183–2200.

Aarhus Convention Compliance Committee. (2017). *Findings and recommendations with regard to communication ACCC/C/2014/102 concerning compliance by Belarus*. Adopted by the Compliance Committee on 18 June 2017 (ECE/MP.PP/C.1/2017/19). https://unece.org/DAM/env/pp/compliance/CC-58/ece.mp.pp.c.1.2017.19.e.pdf. Accessed 01 June 2023.

Ahrens, B. (2017). *The solidarisation of international society: The EU in the global climate change regime* (GLOBUS Research Paper 5/2017). University of Oslo.

Ahrens, B. (2018). *AmbigEUity: The EU and the solidarisation of international society* (PhD thesis). Eberhard Karls Universität Tübingen.

Ahrens, B. (2019). The European Union between solidarist change and pluralist re-enactment. In T. Brems Knudsen & C. Navari (Eds.), *International organisation in the anarchical society: The institutional structure of world order* (pp. 265–292). Palgrave Macmillan.

Ahrens, B., & Diez, T. (2015). Solidarisation and its limits: The EU and the transformation of international society. *Global Discourse, 5*(3), 341–355.

Andresen, S. (2007). The effectiveness of UN environmental institutions. *International Environmental Agreements, 7*, 317–336.

Andrusevych, A., Alge, T., & Konrad, C. (Eds.). (2011). *Case Law of the Aarhus Convention Compliance Committee*. Resource and Analysis Center 'Society and Environment'.

Atapattu, S. (2019). From 'our common future' to Sustainable Development Goals: Evolution of sustainable development under international law. *Wisconsin International Law Journal, 36*(2), 215–246.

Ball, T. (2006). Democracy. In A. Dobson & R. Eckersley (Eds.), *Political theory and the ecological challenge* (pp. 131–147). Cambridge University Press.

Banner, C. (2015). The Aarhus Convention: An overview. In C. Banner (Ed.), *The Aarhus Convention: A guide for UK lawyers* (pp. 1–12). Hart.

Bauer, S., Andresen, S., & Biermann, F. (2012). International bureaucracies. In F. Biermann & P. Pattberg (Eds.), *Global environmental governance reconsidered* (pp. 27–44). MIT Press.

Brady, K. (1998). New convention on access to information and public participation in environmental matters. *Environmental Policy and Law, 28*(2), 69–75.

Briant Carant, J. (2017). Unheard voices: A critical discourse analysis of the Millennium Development Goals' evolution into the Sustainable Development Goals. *Third World Quarterly, 38*(1), 16–41.

Brisman, A. (2013). The violence of silence: Some reflections on access to information, public participation in decisionmaking, and access to justice in matters concerning the environment. *Crime, Law and Social Change, 59*, 291–303.

Churchill, R. R., & Ulfstein, G. (2000). Autonomous institutional arrangements in multilateral environmental agreements: A little-noticed phenomenon in international law. *American Journal of International Law, 94*(4), 623–659.

Coalition for the International Criminal Court. (2023). *United States*. https://www.coalitionfortheicc.org/country/united-states. Accessed 18 May 2023.

Convention on Access to Information, Public Participation in Decision-making and Access to Justice in Environmental Matters. (1998). ECE/CEP/43. https://unece.org/DAM/env/pp/documents/cep43e.pdf. Accessed 09 March 2023.

Day, C. (2015). The Aarhus Convention and NGOs. In C. Banner (Ed.), *The Aarhus Convention: A guide for UK lawyers* (pp. 181–198). Hart.

Death, C., & Gabay, C. (2015). Doing biopolitics differently? Radical potential in the post-2015 MDG and SDG debates. *Globalizations, 12*(4), 597–612.

Dellinger, M. (2012). Ten years of the Aarhus Convention: How procedural democracy is paving the way for substantive change in national and international environmental law. *Colorado Journal of International Environmental Law and Policy, 23*(2), 309–366.

Department for Environment, Food and Rural Affairs. (2020). *Aarhus National Implementation Report 2021*. https://consult.defra.gov.uk/eu/aarhus-national-implementation-report-2021/. Accessed 31 May 2023.

Ebbesson, J., Gaugitsch, H., Jendroska, J., Marshall, F., & Stec, S. (2014). *The Aarhus Convention: An implementation guide*. United Nations.

Etemire, U. (2023a). Public voices and environmental decisions: The Escazú Agreement in comparative perspective. *Transnational Environmental Law, 12*(1), 175–199.

Etemire, U. (2023b). The Escazú Agreement: Public access to environmental information and the goal of a sustainable future. *Journal of Energy and Natural Resources Law, 41*(1), 71–91.

European ECO Forum. (2002,October 20). *European ECO Forum Carignano-Lucca Declaration*. https://cwm.unitar.org/publications/publications/cbl/prtr/pdf/cat1/Carignano_Lucca_declaration.pdf. Accessed 12 October 2023.

European Environmental Bureau. (2022). *NGO statement on Belarus' withdrawal from the Aarhus Convention*. https://eeb.org/library/ngo-statement-on-belarus-withdrawal-from-the-aarhus-convention/. Accessed 01 June 2023.

European Parliament. (2022). *Motion for a resolution on the continuing repression of the democratic opposition and civil society in Belarus* (B9-0508/2022). https://www.europarl.europa.eu/doceo/document/B-9-2022-0508_EN.html. Accessed 01 June 2023.

Falkner, R. (2021). *Environmentalism and global international society*. Cambridge University Press.
Fasoli, E., & McGlone, A. (2018). The non-compliance mechanism under the Aarhus Convention as 'soft' enforcement of international environmental law: Not so soft after all! *Netherlands International Law Review, 65*, 27–53.
Flint, C., & Zhu, C. (2019). The geopolitics of connectivity, cooperation, and hegemonic competition: The Belt and Road Initiative. *Geoforum, 99*, 95–101.
Gilmore, J. (2023). Conceptualising good global statehood: Progressive foreign policy after the populist movement. *International Theory, 15*(1), 79–105.
Hurrell, A. (2007). *On global order*. Oxford University Press.
Jendrośka, J. (2012). Citizens' rights in European environmental law: Stock-taking of key challenges and current developments in relation to public access to information, participation and access to justice. *Journal for European Environmental and Planning Law, 9*(1), 71–90.
Kingston, S., Wang, Z., Alblas, E., Callaghan, M., Foulon, J., Daly, C., & Norris, D. (2022). *Europe's nature governance revolution: Harnessing the shadow of heterarchy* (pp. 1–32). International Environmental Agreements, pre-press online.
Koester, V. (2005). Review of compliance under the Aarhus Convention: A rather unique compliance mechanism. *Journal of European Environmental and Planning Law, 1*, 31–44.
Koester, V. (2007). The Compliance Committee of the Aarhus Convention: An overview of procedures and jurisprudence. *Environmental Policy and Law, 37*(2–3), 83–96.
Koester, V. (2015). The Aarhus Convention compliance mechanism and proceedings before its compliance committee. In C. Banner (Ed.), *The Aarhus Convention: A guide for UK lawyers* (pp. 201–215). Hart.
Krämer, L. (2012). Transnational access to environmental information. *Transnational Environmental Law, 1*(1), 95–104.
Kravchenko, S. (2007). The Aarhus Convention and innovations in compliance with multilateral environmental agreements. *Colorado Journal of International Environmental Law and Policy, 18*, 1–50.
Linklater, A. (2007). Towards a sociology of global morals with an emancipatory intent. *Review of International Studies, 33*, 135–150.
Morgera, E. (2005). An update on the Aarhus Convention and its continued global relevance. *Review of European Community and International Environmental Law, 14*(2), 138–147.
Nye, J. (2023). *Soft power and great-power competition: Shifting sands in the balance of power between the United States and China*. Springer Nature.

Orellana, M. (2016). Governance and the Sustainable Development Goals: The increasing relevance of access rights in Principle 10 of the Rio Declaration. *Review of European, Comparative and International Environmental Law, 25*(1), 50–58.

Peeters, M., & Nobrega, S. (2014). Climate change-related Aarhus conflicts: How successful are procedural rights in EU climate law? *Review of European Community and International Environmental Law, 23*(3), 354–366.

Protocol on Pollutant Release and Transfer Registers to the Convention on Access to Information, Public Participation in Decision-Making and Access to Justice in Environmental Matters. (2003). *United Nations Treaty Series, Volume 2629, A-37770*. https://treaties.un.org/doc/Publication/UNTS/Volume%202629/volume-2629-A-37770.pdf. Accessed 07 May 2023.

Regional Agreement on Access to Information, Public Participation and Justice in Environmental Matters in Latin America and the Caribbean. (2018). *United Nations Treaty Series, 3397, C.N.196.2018*. https://treaties.un.org/Pages/ViewDetails.aspx?src=IND&mtdsg_no=XXVII-18&chapter=27&clang=_en. Accessed 22 May 2023.

Ryall, Á. (2023). A brave new world: The Aarhus Convention in tempestuous times. *Journal of Environmental Law, 35*, 161–166.

Samvel, G. (2020). Non-judicial, advisory, yet impactful? The Aarhus Convention Compliance Committee as a gateway to environmental justice. *Transnational Environmental Law, 9*(2), 211–238.

Special Rapporteur on Environmental Defenders. (2022). *Vision for the mandate: Statement by the Aarhus Convention's Special Rapporteur on Environmental Defenders*. https://unece.org/sites/default/files/2022-11/Vision_for_mandate.pdf?fbclid=IwAR15mf5QmmOEsTMbf9GEifLGB1d15u5vdzpy0NLT4GZCE-IFeMfr5xeWJ2A. Accessed 25 May 2023.

United Nations. (2005). The *secretary-general: Message to second Meeting of the Parties to the Aarhus Convention on access to information, public participation in decision-making and access to justice in environmental matters*. https://unece.org/fileadmin/DAM/env/pp/SG%20message%20final.pdf. Accessed 19 May 2023.

United Nations. (2014, July 2). *The secretary-general: Message to Joint High-Level segment of the meetings of the parties to the convention on access to information, public participation in decision-making and access to justice in environmental matters and its protocol on pollutant release and transfer registers*. Maastricht. https://unece.org/fileadmin/DAM/env/pp/mop5/HLS/SG_statement_Aarhus_MoP5_final.pdf. Accessed 22 May 2023.

United Nations. (2023). *Convention on access to information, public participation in decision-making and access to justice in environmental matters*. Aarhus,

Denmark, 25 June 1998. Guinea-Bissau: Accession. C.N.102.2023.Treaties-XXVII.13 (Depositary Notification). https://treaties.un.org/doc/Publication/CN/2023/CN.102.2023-Eng.pdf. Accessed 22 May 2023.

United Nations Department of Economic and Social Affairs. (2023). *Sustainable Development Goals: History*. https://sdgs.un.org/goals#history. Accessed 22 May 2023.

United Nations Economic Commission for Europe. (2002). *Report of the first Meeting of the Parties* (ECE/MP.PP/2). https://unece.org/fileadmin/DAM/env/documents/2002/pp/ece.mp.pp.2.e.pdf. Accessed 18 May 2023.

United Nations Economic Commission for Europe. (2004a). *Report of the first Meeting of the Parties: Addendum—Decision I/7, review of compliance* (ECE/MP.PP/2/Add.8). https://unece.org/fileadmin/DAM/env/pp/documents/mop1/ece.mp.pp.2.add.8.e.pdf. Accessed 18 May 2023.

United Nations Economic Commission for Europe. (2004b). *Report of the first Meeting of the Parties: Addendum—Decision I/8, reporting requirements* (ECE/MP.PP/2/Add.9). https://unece.org/fileadmin/DAM/env/pp/documents/mop1/ece.mp.pp.2.add.9.e.pdf. Accessed 17 May 2023.

United Nations Economic Commission for Europe. (2004c). *Report of the first Meeting of the Parties: Addendum—Lucca Declaration* (ECE/MP.PP/2/Add.1). https://unece.org/fileadmin/DAM/env/pp/documents/mop1/ece.mp.pp.2.add.1.e.pdf. Accessed 18 May 2023.

United Nations Economic Commission for Europe. (2005a). *Report of the second Meeting of the Parties* (ECE/MP.PP/2005/2). https://unece.org/fileadmin/DAM/env/documents/2005/pp/ece/ece.mp.pp.2005.2.e.pdf. Accessed 19 May 2023.

United Nations Economic Commission for Europe. (2005b). *Report of the second Meeting of the Parties: Addendum—Almaty Declaration* (ECE/MP.PP/2005/2/Add.1). https://unece.org/fileadmin/DAM/env/documents/2005/pp/ece/ece.mp.pp.2005.2.add.1.e.pdf. Accessed 19 May 2023.

United Nations Economic Commission for Europe. (2005c). *Report of the second Meeting of the Parties: Addendum—Decision II/4: Promoting the application of the principles of the Aarhus Convention in international forums*. https://unece.org/fileadmin/DAM/env/documents/2005/pp/ece/ece.mp.pp.2005.2.add.5.e.pdf. Accessed 19 May 2023.

United Nations Economic Commission for Europe. (2005d). *Report of the second Meeting of the Parties: Addendum-Decision II/10: Reporting Requirements*. (ECE/MP.PP/2005/2/Add.14). https://unece.org/fileadmin/DAM/env/documents/2005/pp/ece/ece.mp.pp.2005.2.add.14.e.pdf. Accessed 12 October 2023.

United Nations Economic Commission for Europe. (2007). *Guidance on reporting requirements* (ECE/MP.PP/WG.1/2007/L.4). https://unece.org/fileadmin/DAM/env/documents/2007/pp/ece_mp_pp_wg_1_2007_L_4_e.pdf. Accessed 17 May 2023.
United Nations Economic Commission for Europe. (2008). *Report of the third Meeting of the Parties—Addendum: Riga Declaration* (ECE/MP.PP/2008/2/Add.1). https://unece.org/fileadmin/DAM/env/pp/mop3/ODS/ece_mp_pp_2008_2_add_1_e_Riga.pdf. Accessed 19 May 2023.
United Nations Economic Commission for Europe. (2011a). *Report of the fourth session of the Meeting of the Parties* (ECE/MP.PP/2011/2). https://unece.org/fileadmin/DAM/env/pp/mop4/Documents/Post_Session/ece.mp.pp.2011.2.eng.pdf. Accessed 31 May 2023.
United Nations Economic Commission for Europe. (2011b). *Report of the fourth session of the Meeting of the Parties, Addendum: Chisinau Declaration and decisions adopted by the Meeting of the Parties* (ECE/MP.PP/2011/2/Add.1). https://unece.org/fileadmin/DAM/env/pp/mop4/Documents/ece_mp.pp_2011_2_add.1_eng.pdf. Accessed 17 May 2023.
United Nations Economic Commission for Europe. (2013). *Evaluation of the current functioning and implementation of the convention* (ECE/MP.PP/WG.1/2013/8). http://www.unece.org/fileadmin/DAM/env/pp/wgp/WGP_16/ece.mp.pp.wg.1.2013.8.aev.pdf.Accessed 24 June 2013.
United Nations Economic Commission for Europe. (2014a). *Maastricht Declaration: Transparency as a driving force for environmental democracy* (ECE/MP.PP/2014/27/Add.1). https://unece.org/fileadmin/DAM/env/pp/mop5/HLS/ece.mp.pp.2014.27.Add.1_e.pdf. Accessed 22 May 2023.
United Nations Economic Commission for Europe. (2014b). *Maastricht recommendations on promoting effective public participation in decision-making in environmental matters* (ECE/MP.PP/2014/2/Add.2). https://unece.org/fileadmin/DAM/env/pp/mop5/Documents/Post_session_docs/ece_mp.pp_2014_2_add.2_eng.pdf. Accessed 22 May 2023.
United Nations Economic Commission for Europe. (2014c). *Report of the fifth session of the Meeting of the Parties* (ECE/MP.PP/2014/2). https://unece.org/fileadmin/DAM/env/pp/mop5/Documents/Post_session_docs/ece_mp.pp_2014_2_eng.pdf. Accessed 22 May 2023.
United Nations Economic Commission for Europe. (2017a). *Budva Declaration on environmental democracy for our sustainable future* (ECE/MP.PP/2017/16/Add.1). https://unece.org/fileadmin/DAM/env/pp/mop6/HLS/ECE_MP.PP_2017_16_Add.1-ECE_MP.PRTR_2017_2_Add.1_E.pdf. Accessed 23 May 2023.
United Nations Economic Commission for Europe. (2017b). *Report of the sixth session of the Meeting of the Parties* (ECE/MP.PP/2017/

2). https://unece.org/fileadmin/DAM/env/pp/mop6/English/ECE_MP.PP_2017_2_E.pdf. Accessed 17 May 2023.

United Nations Economic Commission for Europe. (2019). *Guide to the Aarhus Convention Compliance Committee*. United Nations Economic Commission for Europe. https://unece.org/DAM/env/pp/Publications/Guide_to_the_Compliance_Committee__second_edition__2019_/English/Guide_to_the_Aarhus_Convention_Compliance_Committee__2019.pdf. Accessed 26 May 2023.

United Nations Economic Commission for Europe. (2021). *Synthesis report on the status of implementation of the Convention* (ECE/MP.PP/2021/6). https://unece.org/sites/default/files/2021-10/ECE_MP.PP_2021_6_E.pdf. Accessed 17 May 2023.

United Nations Economic Commission for Europe. (2022a). *Geneva Declaration on environmental democracy for sustainable, inclusive and resilient development* (ECE/MP.PP/2021/16/Add.1). https://unece.org/sites/default/files/2022-02/ECE_MP.PP_2021_16_Add.1_E_6.pdf. Accessed 25 May 2023.

United Nations Economic Commission for Europe. (2022b). *Report of the seventh session of the Meeting of the Parties* (ECE/MP.PP/2021/2). https://unece.org/sites/default/files/2022-07/ECE_MP.PP_2021_2_E.pdf. Accessed 17 May 2023.

United Nations Economic Commission for Europe. (2022c). *The Compliance Committee in figures*. https://unece.org/sites/default/files/2022-02/ACCC_Infographics.pdf. Accessed 30 May 2023.

United Nations Economic Commission for Europe. (2023a). *Aarhus Convention national implementation reports: Reporting mechanism*. https://aarhusclearinghouse.unece.org/national-reports/reports. Accessed 17 May 2023.

United Nations Economic Commission for Europe. (2023b) *Bureau*. https://unece.org/env/pp/aarhus-convention-bureau. Accessed 25 May 2023.

United Nations Economic Commission for Europe. (2023c). *The committee's cases*. https://unece.org/environment-policy/public-participation/aarhus-convention/compliance-committee. Accessed 30 May 2023.

United Nations Economic Commission for Europe. (2023d). *Twenty-seventh meeting of the Working Group of the Parties to the Aarhus Convention*. https://unece.org/environmental-policy/events/twenty-seventh-meeting-working-group-parties-aarhus-convention. Accessed 25 May 2023.

United Nations Environment Programme. (2015). *Bali guideline implementation guide*. https://www.unep.org/resources/publication/bali-guideline-implementation-guide. Accessed 22 May 2023.

United States Department of State. (2002, May 6). *International Criminal Court: Letter to UN Secretary General Kofi Annan*. Washington, DC. https://2001-2009.state.gov/r/pa/prs/ps/2002/9968.htm. Accessed 18 May 2023.

United Nations Treaty Collection. (2023). *Convention on access to information, public participation in decision-making and access to justice in environmental matters*. Status of Ratification. https://treaties.un.org/Pages/ViewDetails.aspx?src=IND&mtdsg_no=XXVII-13&chapter=27&clang=_en. Accessed 23 May 2023.

Vienna Convention on the Law of Treaties. (1969). *United Nations Treaty Series, 1155, 331*. https://legal.un.org/ilc/texts/instruments/english/conventions/1_1_1969.pdf. Accessed 06 April 2023.

Warren, M. (2017). *The all-affected interests principle in democratic theory and practice* (Working paper). Institute for Advanced Studies. http://aei.pitt.edu/93142/1/Warren_-_All_Affected_Interests_Principle.pdf. Accessed 08 May 2023.

Wates, J. (2005). The Aarhus Convention: A driving force for environmental democracy. *Journal of European Environmental and Planning Law, 1*, 2–11.

Wates, J. (2011a). *Statement by Mr Jeremy Wates, Secretary-General of EEB, European Eco forum, and the former Secretary of the Aarhus Convention*. Meeting of the Parties to the Convention on Access to Information, Public Participation in Decision-making and Access to Justice in Environmental Matters. Fourth session. https://unece.org/fileadmin/DAM/env/pp/mop4/Statements/Statement_by_Jeremy_Wates_EEB_ECOForum.pdf. Accessed 22 May 2023.

Wates, J. (2011b). The future of the Aarhus Convention: Perspectives arising from the third session of the Meeting of the Parties. In M. Pallemaerts (Ed.), *The Aarhus Convention at ten: Interactions and tensions between conventional international law and EU environmental law* (pp. 383–415). Europa Law.

Weaver, D. (2015), *The Aarhus Convention: Towards a cosmopolitan international environmental politics* (Unpublished PhD thesis). Keele University. https://eprints.keele.ac.uk/id/eprint/2310/1/Weaverphd2015.pdf. Accessed 12 April 2023.

Weber, T. (2022). Are climate activists protected by the Aarhus Convention? A note on Article 3(8) Aarhus Convention and the new Rapid Response Mechanism for environmental defenders. *Review of European, Comparative and International Environmental Law, 32*, 67–76.

World Commission on Environment and Development. (1987). *Our common future*. Oxford University Press.

Yu, S. (2019). The Belt and Road Initiative: Modernity, geopolitics and the developing global order. *Asian Affairs, 50*(2), 187–201.

Zhao, Y., & Butcher, B. (2022). Coming to terms with public participation in decision making: Balancing clarity and impact in the Aarhus Convention. *Review of European, Comparative and International Environmental Law, 31*, 210–221.

CHAPTER 6

Conclusion: Towards Solidarisation

Introduction

Let us return to where we began. I opened the book with the admittedly quaint idea that international environmental agreements were akin to cultivated flowering plants. I cast them as being propagated, as germinating, as sometimes growing and reproducing. I also cast them as contributing to the health and well-being of their broader ecosystem. And I paid attention to their deliberate cultivation by their "gardeners", the states that consented to their emergence and growth.

After setting an English School framework offering three competing and complementary imaginaries of IR—realism, rationalism and revolutionism (Wight, 1991)—I took my chosen agreement, a particularly beautiful flower in the multilateral garden. The Aarhus Convention was purposively selected because, in its first quarter century, it had been praised as a democratising pioneer in international environmental politics and law, by the highest players on the world stage, and by the most accomplished lawyers. Yet only scant engagement had ensued with Aarhus' international ethical implications. Before this book, I had only begun to scratch the surface of Aarhus' ethical traction (Weaver, 2018). Far more was needed. The question remained as to exactly what ramifications the Convention yielded for human-oriented, environmentally minded ethical progress in a world that continued to be prevailed upon by states. That was the task, and that is what I hope to have achieved.

D. Weaver, *The Aarhus Convention*, Environmental Politics and Theory,
https://doi.org/10.1007/978-3-031-43536-2_6

To engage Aarhus' discretely ethical rationales and realities, I adopted ethically attuned English School lenses to critically explore Aarhus' *propagation*, namely its preparations, negotiations and emergence, and the contexts leading there; its *germination*, taking effort to excavate the ethics of its trinity of procedural rights; and its *growth*, paying attention to the development of its organisational infrastructure. My findings speak to a fresh contribution by contemporary English School thinkers, solidarisation. As we come to a close, the task is to discuss the implications for English School understandings of environmental multilateralism. It is appropriate, in this regard, to comment in turn on Aarhus' relevance for the three traditions (Wight, 1991), and to close with solidarisation.

Realism

Only limited indications of realism were gauged from the exploration of Aarhus that ensued. But where they *were*, the risks were significant. The critical risk is that of a realist regression in Aarhus' implementation, at a time when hard geopolitics returned to Europe. It is disheartening for me to have to cast Aarhus under the shadow of calculating realist arrangements in which states' interactions are "sufficient to make the behaviour of each a necessary element in the calculations of the other" (Bull, 2002: 9–10). How counterintuitive it feels, to have to draw conclusions relating to enmity between the world's sovereigns in IR. Nonetheless, to draw these conclusions is necessary.

We witnessed rare but disquieting instances of state survival interests being pursued over the mandates of international cooperation, environmental protection and human rights. During Aarhus' propagation, we saw that the Russian Federation's refusal to sign Aarhus was likely attributable to the environmental risks posed by its military assets: risks exposed by someone who would now be termed, in the current lexicon, an environmental defender. The Russian Federation will have calculated that its national interests—its military and security interests—would have been hindered rather than helped by its membership.

Of more pressing and immediate concern was Belarus' withdrawal from the Aarhus Convention, following the MOP having had no option other than to initiate a rapid response mechanism and an associated Special Rapporteur on Environmental Defenders. This I deem Aarhus' most troubling development, and the greatest risk to Aarhus' growth. UNECE, Aarhus' Parties and civil society must rally together to ensure that such an instance of regressive unilateralism remains isolated, and sets no precedent for actors that might be under political or economic pressure to distance themselves from their multilateral environmental and democratic commitments.

Moreover, on the issue of the rapid response mechanism, I never perceived any hint, from the archives and from my doctoral fieldwork, that anything more than Article 3.8's prevention of "penalisation, persecution and harassment" (Aarhus Convention 1998) would be needed during the Convention's implementation. That this mechanism was required in the first place indicates the very real risks facing environmental defenders, not least inside the UNECE region. Such risks are palpable as Aarhus marks the twenty-fifth anniversary of its opening for signature.

The risks are illuminated brightly in the context of the Russian Federation's war in Ukraine, which entailed an annexation of territory in 2014 and a full-scale war, spanning all of Ukraine and yielding global implications, in 2022. A few weeks before this book was sent to the press in 2023, hard geopolitics hit the Aarhus community with the collapse of the Kakhovka Dam, a Ukrainian hydroelectric dam that had been taken by the Russian Federation in 2022, and which had remained in Russian custody until its collapse. The environmental harm that was sustained, and which may yet still be sustained, is significant.. The dam had once helped to maintain the Dnipro River reservoir that provides cooling water to Zaporizhzhia nuclear power station, Europe's biggest nuclear power station and, at the time of writing, a pivot around which warring forces appear to oscillate.

I find it difficult to separate Belarus' withdrawal from the Aarhus Convention, from its implication in this sombre return of hard geopolitics. Such a return implies the risk of ethical relapse: towards state survivalism, and towards militarism, replete as they are with risks of environmental harm. I also perceive a risk of ethical relapse from the human and environmental rights progress achieved by current and former Parties. I appreciate that UNECE is providing, in the very best of faith, as much support as it possibly can in the ongoing war. Ukraine's active solicitation of assistance from the Compliance Committee, and the latter's support to help Ukraine manage its compliance during warfare, heralds a milestone in Aarhus' lifespan. I suspect that some manual of guidance or other will eventually emerge from this compliance activity, and I suspect such a manual will formalise the procedures needed for the safe and effective exercise of environmental rights and responsibilities during wartime (see

Ryall, 2023). Yet this laudable work is at risk of yielding pyrrhic victories: retrospective instances of positive progress that achieved too little, too late, with environmental harm and human tragedy already having been sustained. To ponder such regression is difficult. Yet these risks should not be deemed irreparable. They should not dim the brilliance of Aarhus' achievements in its first quarter century. Were I to frame these thoughts in English School logic, I would sit myself near James Mayall (2000: 157), who once aptly noted that "we have no realistic alternative than to approach the future with caution, but also with hope".

Revolutionism

It is also the case that only limited indications of revolutionism were gauged. What I *did* observe, in Aarhus' propagation, germination and growth, was a tangible role played by what English School thinkers would term world society actors: humans and their collectivities constituting the nonstate, civil societal domain. World society actors were seen to have enjoyed relative parity of actorhood. On some occasions, a very slim degree of functional hybridity arose. I recall the joint proposals by civil society and state delegates during Aarhus' negotiations. Such co-creation only appears to have intensified in Aarhus' subsequent growth period, given the powerful and moving articulations of voice by civil society actors . Human voice continues to be articulated and heard in relative parity with that of state. Such a co-creative modality, underway in the Convention, peppered as it is with a strictly limited functional hybridity, is as close as Aarhus gets to revolutionism.

But I must make it clear that the book observed humans' *empowerment*, not *emancipation*. There was no unshackling of world society from sovereignty. Actual revolutionism—actual instances of human emancipation from the chains repressing human freedom—was unapparent. Humans were empowered within statist international society arrangements. As will be reinforced in the next section, I gauged a cosmopolitan impulse that drove Aarhus' progress *inside* existing power arrangements. Cosmopolis never materialised.

It is in this context that the great canon of human and environmental rights preceding Aarhus' emergence—whilst laudable in setting the context for Aarhus to be successfully propagated—never outgrew the states consenting to it. Sovereignty was never usurped. Aarhus was negotiated under the auspices of, and is administered within, power frameworks that themselves hinge upon sovereignty.

Similarly, the voice of civil society, articulated as it was so well during Aarhus' lifespan to date, was unprecedented when Aarhus emerged. We seldom observe such capacity for humans to *receive*—and articulate—such well amplified voice in multilateralism. To this extent, Aarhus was a pioneer, significantly "tuning up" the human capacity to hear and be heard in the international arena, for environmental purposes. Yet I stress "*receive*" because the receipt of these rights is central to why such scant evidence of revolutionism surfaced. World society did not declare its own rights. It did not develop its own organic capacities to articulate itself and its environmental concerns. World society *received* those rights from Aarhus' consenting states (see Buzan, 2018). States bestowed upon humans these rights. States granted humans and their collectivities the enhanced capacities to influence environmental decision-making. The human manifests itself here as a state's beneficiary, even in the context of such a radical agreement as Aarhus enabling direct rights-claiming against states by humans.

This implies residual and persistent power imbalances between sovereign rights-*providers* and their subjects. The territory of IR, here, is undeniably one upon which sovereigns continue to govern their human and environmental realms. Yet, as will be reinforced in the next section, the *modus operandi* of sovereignty, in the international society delineated by Aarhus' Parties, is itself subjected to ethical maturation.

The impossibility of revolutionism is not something that we should lament. Never was it my intention to go hunting for post-sovereign, transformative potentialities. Such hunting carries risks. However tempting it might be to prophesy a world breaking free from sovereignty, whose humanity has unchained itself, the risks are significant. They relate to the destabilising impact that emancipation would have on international order. As I write, never (in my lifetime) has there been a more critical moment at which the need for stable international order has manifested itself so acutely. Now is not the time for IR's thinkers and practitioners to dabble in post-sovereign fantasy. Humanity and the environment need stable international order. Without it, human rights, and the rich natural environment that my readers no doubt cherish, will elude us like sand through our fingers.

To attempt to "free" humans from the very rules-based international order needed to protect them would be to undo centuries' efforts to *rationalise* our political world and render it as nonviolent as possible, given the lack of a supranational authority with which to curtail states' harmfulness. Cosmopolis is simply not the answer. What is essential is a responsible, rationalising approach to multilateralism, which *empowers* humans without *emancipating* them, and which contains the limits of progress within states' thresholds of tolerance. It is to this final closing position that I turn.

Rationalism: Towards Solidarisation

We witnessed, in this book, a deliberate attempt by a limited number of states to converge on a shared issue pertaining not only to human and environmental rights, but to the garnering of a richer, more participatory form of democratic governance. The issue concerned the difficulties in codifying an internationally recognised substantive right to an adequately clean and healthy environment. It consequently related to the need to entrench a suite of procedural rights to information access, public participation and access to justice in environmental matters. Without the codification of such multilaterally agreed procedural rights, effective pursuit of substantive environmental rights—by states in concert with humans—would be chimeric and nebulous.

In this deliberate convergence, consenting sovereign states formed an international society delineated by their Aarhus Party status. These regional, issue-defined international society arrangements were characterised by stricter standards of civilisation (Gong, 1984): stricter rules that needed following, in order for Parties to maintain their legitimacy as sovereigns in the eyes of their populations, state counterparts and transnational civil society. I deem Aarhus' international society arrangements greener, more democratic and more "humankind", in enabling humans and their associations to more effectively express themselves: not for the vanities of gratification, or for "freedom" per se, but for the discrete purposes of securing—for themselves, their societies and their states—a cleaner, healthier environment. The procedural rationalising that is conducted, in these limited international society arrangements, contributes to the pursuit of substantive environmental protection within and between Party jurisdictions.

In constituting this greener, more democratic international society, states collectively invested their sovereignty, placing aside their sometimes chasmic differences for the greater good of the *inter*national interest, of humanity and nonhuman nature. Environmental dialogue ensued between erstwhile Cold War antagonists at, inter alia, Bellagio and Kiev. It intensified during Environment for Europe. As states' convergence strengthened, we witnessed them collectively saying "no more" to the violent quelling of environmental activism in the era of *Glasnost* and *Perestroika*. Former Warsaw Pact members converged with market and political liberal democracies: once adversaries caught in the abyss of realist calculation. These sometime antagonists mindfully converged to propagate, germinate and grow something beautiful in the garden of multilateralism. The latter was rationalised just that little more democratically, green-mindedly and humanely, by the Aarhus Convention.

Solidarist achievements were garnered. I made the case, throughout the book, that a moral cosmopolitan impulse of "humankindness" breathed life into these new international society arrangements. The cosmopolitan impulse drove Aarhus' concerns for human empowerment; for amplifying human voice; for institutionalising the ancient all-affected principle; and for codifying such a degree of state-public co-creation as is seldom witnessed in the traditional powerhouses of international law and diplomacy.

Chapter 3's coverage of Aarhus' *propagation* saw world society actors infusing Aarhus' text with more demanding solidarist intent. We saw proposals pitched by world society actors in concert with state delegates. We witnessed a sustained and concerted attempt by state and nonstate actors to codify human and environmental rights that would apply to all persons, and which would enable some modicum of socioeconomic development, across the Global North and South, whilst heeding the natural environment's needs.

Chapter 4's treatment of the Convention's *germination* gauged, from its trinity of procedural rights, a codified amplification of human voice; binding commitments to state-public dialogue; a spirit of co-creation; a transition from "need to know" to "right to know" information governance; and a concerted materialisation of the all-affected principle via the due account provision. An attempt at overcoming the violence of silence (Brisman, 2013) was made, and it shined brightly given the implication of the third pillar—as well as the compliance mechanism's public trigger—in a turn towards green *actio popularis*.

Chapter 5's exploration of Aarhus' *growth*, in terms of the development of its organisational infrastructure, identified a typically warm disposition to civil society, which had been garnered during the propagation period and which never dissipated. Such inclusiveness was evident across the strategic, operational and tactical tiers, but was most acutely evident at the tactical level, the compliance mechanism, which bestows some degree of legal actorhood on humans and their associations, and whose quasi-judicial practices bruised the more pluralist letter of the law, given the mechanism's enablement of citizen enforcement, its jurisprudential character, and its commitment to equity, diversity and inclusion in the Compliance Committee's composition.

Yet throughout this book, the fostering of radical solidarism was seen to be tempered by a counterbalancing in favour of sovereignty. The pluralist power frameworks, in which international law and diplomacy are played out, were reaffirmed and strengthened, rather than undermined and attenuated. The counterbalancing was weighted in favour of states' sovereign prerogatives: for a state to express reservation pursuant to classical international law; for a state to protect and preserve its interests, in particular its national security and military priorities; for a state to pursue the stabilisation of international order, in a bid to avoid violence in a world lacking Leviathan. "While inspired by lofty ideals", Wates (2011: 414) once wrote, the Aarhus Convention is thus "a pragmatic compromise, like the outcome of most international negotiating processes".

Where solidarism was sought, in Aarhus' first quarter century, it was accompanied by clear iterations of sovereignty. Underpinning this *persistence of*, and *insistence upon*, sovereignty was the tenacity of state consent, which I characterised as states' unwavering ability and willingness to enter and withdraw from the diplomatic arena at their—and only their—discretion. The possibility of compelling states to act in a given way, in the arena of environmental multilateralism, is very slim. After all, "Activism in world politics is voluntary rather than obligatory" (Jackson, 2003: 417). Aarhus' international society arrangements are without any doubt anarchical (Bull, 2002), because the consenting states, which partake in these international relations of environmental accountability, do so voluntarily and with conscious intent. They deliberately invest their sovereignty together, in the hope of a "return on investment". But they do this, not for the ends of selfish survivalism, but to rationalise their mutual environmental and democratic relations, for the good of their *collective* populations, territories and environments. This is the mark of an international society arrangement: the setting aside of differences, in order to converge upon issues of shared concern for the benefit not only of the individual sovereign state, but also of the *society itself*.

Such counterbalancing in favour of sovereignty ensued during the propagation period. Aarhus' antecedent contexts were intrinsically statist. States kept a firm grip on the primary institutionalisation of human and environmental rights, despite the latter's ethical seeds in the first instance being unpacked and scattered in IR's landscape by world society actors (Falkner & Buzan, 2019). Moreover, it was demonstrated that states' key priority, during Aarhus' propagation, was the deliberate avoidance of another world war. The need for international order was reaffirmed in the often "soft law" aspirations that reflected the need to balance (a) the pluralist requirement for preserving diversity and coexistence in a world recovering from total war and (b) the solidarist need to apprehend, on a *global scale*, the *global risks* materialising in a world that was quickening and shrinking, and whose borders were increasingly unable to protect states and humans from harms that trampled through the passport gate.

This need to balance the stabilisation of international order and the institutionalisation of justice norms helps us to understand the balancing acts, during propagation, between the rhetorics of human and environmental rights, of avoiding war, and of pursuing extractive agendas of industrial development, awkwardly appended to which were unconvincing allusions to industrial hygiene and living conditions. The need to balance the stabilisation of international order and the institutionalisation of justice norms helps us understand why so much of the antecedent context was nonbinding: states were gingerly stepping onto the solidarist terrain of postwar IR at a pace, and with an intensity, acceptable to them. Only tentative steps towards the codification of solidarist norms were taken, and as cautiously as possible, in order to avoid compromising the "safety scaffolding" of international order. Any cosmopolitan intent that was sown was necessarily realistic and incremental (Beck & Sznaider, 2006).

Such counterbalancing in favour of sovereignty was also evident in the germination of Aarhus' trinity of procedural rights. Exemptions were written into it, in order to ensure that Parties retained ultimate sovereignty *over* the provisions. The public interest test for disclosure, and the need to extract disclosable content from exempted material, of course constitute counterchecks against sovereign tyranny. But those exemptions, as I argued, are nonetheless inevitable in an international environmental agreement that places such a burden of positive obligation on its Parties. The latter need to be satisfied that, in discharging their obligations, their primordial communitarian responsibilities—to protect people, territory, societal order, and their *particular* forms of internal authority and external legitimacy—would not be impeded.

It is in this regard that I observed, throughout the book, sovereignty neither being usurped nor undermined by Aarhus' ethical developments (see Dannenmaier, 2008: 62). And counterbalancing as iteration of "sovereignty *over*" was not limited, of course, to Aarhus' exemptions. It was identified in the provision for the public concerned to be identified, and also in states' residual capacity to determine locus standi. It was also identified in the firm cementing of Aarhus' organisational infrastructure with mainline statist multilateralism. The strategic and operational tiers complied almost entirely with state-centric multilateralism. Scarce indications arose of any inclination away from traditional statist power frameworks. Aarhus' "working parts" were the ones expected from such an international environmental agreement. States' laggardness, and the doggedness of sovereignty, weighted the counterbalancing down further. I recall at this juncture the NIR difficulties, plus the rapid response mechanism. I can only logically attribute the latter to the presence of unsatisfactory implementation regimes *inside* Party jurisdictions.

A *cautious incrementalism* was therefore observed. State consent was used not to fortify the ego and arrogance of statehood, but to prevent the undermining of sovereignty, so that sovereignty could be collectively invested for other-oriented endeavours. A tool for achieving this, in the multilateral arena, is the iteration of the sovereign prerogative: the expression, verbally and in writing, of the sanctity of a state's sovereignty, and of its prerogative to enter or withdraw from multilateral proceedings at will. Such iterations reinforce and perpetuate diversity in IR, a pluralist mainstay, because they offer a buffer against the *universalising* imposition of norms on what remain inherently *particular* jurisdictions.

I do not claim that such iterations of sovereignty *prevent* attempted value universalisation. Rather, I regard them as buffering states' sovereign prerogatives, in such a way that their delegates, professional statespersons, feel sufficiently comfortable and secure taking the "little steps" towards a moral cosmopolitan ethics, which are needed to achieve human-oriented solidarist progress. In this regard, any solidarist progress will necessarily remain embedded in, and representative of, only those standards of conduct that "the statespeople involved could reasonably be expected to acknowledge and abide by" (Jackson, 2003: 406).

Solidarist traction will, I conclude, inevitably be contained within, and confined by, the "safety scaffolding" of international order whose guarantor is pluralism and whose talisman is sovereignty. The solidarism that I observed in this book derives from what Williams (termed "a prior and necessary international society predicated upon the delivery of order" [Williams, 2013: 135–136]). Stabilisation of international order, by the responsible practices of sovereignty, is, I argue, a prerequisite for solidarist progress. Yet I stress "responsible" (Aalberts & Werner, 2011: 2200) because sovereignty itself is ethically agitated and matured, given that it must meet stricter standards to be deemed legitimate. As such, solidarism will be an elusive mystery unless it is pursued within the accepted thresholds of a *responsible* sovereignty (Weinert, 2007), and to the extent that it avoids bruising the sovereign prerogative too harshly. From this vantage point, it is indeed, as Hurrell (2007: 292) suggested, "wholly wrong to see pluralism as belonging solely to a vanished Westphalian world". And it may well be the case that such a communitarian approach as pluralism offers a "path to cosmopolitanism" (Shapcott, 2001: 31), so long as that path is taken as cautiously as need be, to avert the destabilisation of international order.

To distil my findings in a closing case, I propose that solidarist progress will only be as transformative as states allow. Sovereignty sets the parameters in which solidarism is confined. The trajectory that Aarhus took, towards a cosmopolitan "humankindness" in environmental contexts, is anchored within the bounds of sovereignty, and thus remains irremovably positioned in the domain of *state-centric*, yet *human-oriented* solidarism. Such a finding is in line with Ahrens' position that "change is much more likely to be moderate or incremental because it is restricted by the previously existing order" (Ahrens, 2019: 268). My findings resonate with solidarisation. And it is with a final definition of solidarisation that I finish: Solidarisation is a process in which sovereignty (a) undergoes ethical maturation by solidarist endeavours, thus becoming more responsible and other-oriented, whilst (b) itself delimiting those solidarist endeavours, to prevent its own usurpation. Solidarisation negates the likelihood of post-sovereign transformation, whilst ensuring the feasibility of a cautiously incremental ethical progress in IR, in order to avoid destabilising international order.

Bibliography

Aalberts, T., & Werner, W. (2011). Mobilising uncertainty and the making of responsible sovereigns. *Review of International Studies, 37*(5), 2183–2200.

Ahrens, B. (2019). The European Union between solidarist change and pluralist re-enactment. In T. Brems Knudsen & C. Navari (Eds.), *International organisation in the anarchical society: The institutional structure of world order* (265–292). Palgrave Macmillan.

Beck, U., & Sznaider, N. (2006). Unpacking cosmopolitanism for the social sciences: A research agenda. *British Journal of Sociology, 57*(1), 1–23.

Brisman, A. (2013). The violence of silence: Some reflections on access to information, public participation in decisionmaking, and access to justice in matters concerning the environment. *Crime, Law and Social Change, 59*, 291–303.

Bull, H. (2002). *The anarchical society: A study of order in world politics* (3rd ed.). Palgrave.

Buzan, B. (2018). Revisiting world society. *International Politics, 55*, 125–140.

Dannenmaier, E. (2008). A European commitment to environmental citizenship: Article 3.7 of the Aarhus Convention and public participation in international forums. *Yearbook of International Environmental Law, 18*(1), 32–64.

Falkner, R., & Buzan, B. (2019). The emergence of environmental stewardship as a primary institution of global international society. *European Journal of International Relations, 25*(1), 131–155.

Gong, G. (1984). *The standard of civilisation in international society*. Clarendon Press.

Hurrell, A. (2007). *On global order*. Oxford University Press.

Jackson, R. (2003). *The global covenant: human conduct in a world of states*. Oxford University Press.

Klinke, A. (2012). Democratising regional environmental governance: Public deliberation and participation in transboundary ecoregions. *Global Environmental Politics, 12*(3), 79–99.

Mayall, J. (2000). *World politics: Progress and its limits*. Polity Press.

Ryall, Á. (2023). A brave new world: the Aarhus Convention in tempestuous times. *Journal of Environmental Law, 35* 161–166.

Shapcott, R. (2001). *Justice, community and dialogue in international relations*. Cambridge University Press.

United Nations Economic Commission for Europe. (2023). *Twenty-seventh meeting of the Working Group of the Parties to the Aarhus Convention*. https://unece.org/environmental-policy/events/twenty-seventh-meeting-working-group-parties-aarhus-convention. Accessed 25 May 2023.

Wates, J. (2011). The future of the Aarhus Convention: Perspectives arising from the third session of the meeting of the parties. In M. Pallemaerts (Ed.), *The Aarhus Convention at ten: Interactions and tensions between conventional international law and EU environmental law* (pp. 383–415). Europa Law.

Weaver, D. (2018). The Aarhus Convention and process cosmopolitanism. *International Environmental Agreements, 18*(2), 199–213.

Weinert, M. S. (2007). Bridging the human rights-sovereignty divide: Theoretical foundations of a democratic sovereignty. *Human Rights Review, January-March, 2007*, 5–32.

Wight, M. (1991). *International theory: The three traditions*. Leicester University Press.

Williams, J. (2013). The international society—World society distinction. In C. Navari & D. Green (Eds.), *Guide to the English school in international studies* (pp. 127–142). Wiley.

Index

A
Aarhus Convention
accession to the Convention, 152
Ad Hoc Working Group, 80
Almaty Guidelines on promoting Aarhus' principles in international forums, 149
annexes, 8
bureau, 12, 157
citizenship, nationality and domicile, 4, 7, 150, 166
compliance mechanism, 8, 12, 142, 154, 159, 160, 163, 165–167
Distinction between *public* and *public concerned*, 109
due process, 165
entrance into force, 8, 85–89, 154, 162
Environmental NGOs Coalition, 80
geopolitical context, 11, 49, 50, 69, 70, 76, 77, 87
germination: procedural trinity, 11, 41, 80, 107, 132, 141, 148, 150, 151, 182, 187
GMO Amendment, 119, 157
growth: development of organisational infrastructure, 141
internationalisation and pollination overseas, 151
legal and diplomatic context, 11, 50, 127
legal review and remedies, 131
letter of the law versus spirit of the law, 150, 160
Meeting of the Parties, 8, 12, 165
fifth MOP, Maastricht, the Netherlands, 152
first MOP, Lucca, Italy, 144, 148, 157, 160, 161
fourth MOP, Chisinau, Moldova, 145, 150
second MOP, Almaty, Kazakhstan, 119, 145, 148
seventh MOP, Geneva, Switzerland, 146, 154–156, 164

D. Weaver, *The Aarhus Convention*, Environmental Politics and Theory,
https://doi.org/10.1007/978-3-031-43536-2

sixth MOP, Budva, Montenegro, 145, 146, 153
third MOP, Riga, Latvia, 150
opening for signature, 77, 88, 181
operational tier, 12, 157, 158, 188
organisational infrastructure, 12, 30, 142, 168, 169, 180, 186, 188
penalisation, persecution and harassment, 154, 155, 181
postwar pursuit of international order, 30, 41
preamble, 5–7, 60, 79
procedural trinity, 5, 105, 107, 110, 132
propagation: preparation, negotiation and emergence, 10, 41, 50, 69, 70, 72–75, 77–79, 83, 84, 87–89, 132, 141, 148, 167, 180, 182, 185, 187
PRTR Protocol, 116, 143
public interest test, 113, 118, 133, 187
public trigger, 166, 185
rapid response mechanism, 12, 118, 142, 154–156, 168, 180, 188
ratification, 8, 148
risk of regression, 154, 156
secretariat, 8, 12, 154, 157, 159
Special Rapporteur on Environmental Defenders, 154, 155, 180
specific activities, 119–121, 123, 126, 127
strategic tier, 12, 152
tactical tier, 12, 142, 143, 168, 186
task forces, 12, 157, 158
Working Group of the Parties, 12, 157, 181, 182
Aberystwyth school of critical IR and security theory, 26
Aberystwyth University, 19
Access to information, 7, 63, 108, 111, 128
Access to justice, 4–8, 62, 66, 74, 78–80, 85, 105, 108–110, 127, 129, 130, 150, 184
Accountability, 4, 6, 30, 38, 41, 63, 71, 76, 79, 80, 87, 107, 111, 113, 115, 117, 118, 121, 127, 150, 186
Actio popularis, 12, 109, 127, 129, 131, 133, 142, 166, 167, 185
All-affected principle, 11, 62, 108, 110, 115, 117, 119, 122, 125, 132, 149, 185
Alliances, 23
Anarchy, 11, 19, 22, 23, 30–33, 40
Animus dominandi, 24
Authority, 2, 19, 32, 106, 112, 113, 118, 121, 124, 128, 131, 155, 184, 188

B

Balance of power
bipolarity, 70
Belarus
civil society organization *Ecohome*, 155
implication in Russian invasion of Ukraine, 30, 155
Reservation regarding Special Rapporteur on Environmental Defenders, 142
withdrawal from the Aarhus Convention, 180, 181
Bellagio Conference on US and Soviet Environmental Protection Institutions, 73
Biosphere, 25, 58
Blocs, 23, 49, 70
Borders
as power containers, 24

bordering processes, 23
boundedness, 26, 34, 36
Brundtland Report. *See* World Commission on Environment and Development (WCED)

C

Canada, 84, 161
Cautious incrementalism, 3, 68, 188
Charter of the United Nations, 30
Checks and balances, 11, 122
China, 156
Belt and Road Initiative, 156, 157
Civil society, 2, 7, 30, 40, 50, 64, 71–73, 75, 76, 80, 81, 87, 89, 110, 141, 147, 148, 153, 155, 156, 158, 159, 167, 180, 182–184, 186
Climate change, 25, 38
Co-creation, 6, 11, 12, 19, 39, 41, 59, 62, 65, 67, 69, 78, 81, 83, 88, 106–108, 110, 117, 119, 121, 122, 125, 126, 132, 148, 156, 168, 169, 182, 185
Coexistence, 9–11, 13, 22, 33, 35, 40, 50–52, 68, 88, 187
Cold War
ideological balance of power, 70
'long peace', 70
Nuclear balance of power, 70
Communitarianism, 34, 37, 41, 189
Conditionality, 87
Conference on Security and Cooperation in Europe (CSCE), 75, 76
Meeting on the Protection of the Environment, 75
Convention on Biological Diversity (CBD), 65
Convention on Environmental Impact Assessment in a Transboundary Context, 61
Convention on Long-Range Transboundary Air Pollution, 61
Convention on the International Right of Correction, 54
Cosmopolis, 10, 20, 21, 26, 27, 31, 40, 182, 184
Cosmopolitanism
cosmopolitan citizenship, 28, 29
cosmopolitan impulse, 6, 8, 11, 26, 31, 38, 39, 59, 69, 78, 88, 89, 105, 110, 113, 117, 119, 121, 125, 129, 130, 133, 141, 148, 157, 167–169, 182, 185
'humankindness', 7, 9, 12, 41, 63, 89, 141, 157, 169, 185, 189
moral cosmopolitanism, 33, 37, 41, 108, 109, 117, 125, 133, 141
of 'little steps', 39, 188
realistic cosmopolitanism, 9, 39
runaway or unbridled cosmopolitanis, 126, 165
transformation, 39
Council of Europe, 55, 56
Council of Europe Convention on Access to Official Documents, 55

D

Democracy
liberal democracy, 71, 185
market liberalism, 29
political liberalism, 28, 71
Democratisation, 4, 11, 50, 69, 72, 74–77, 80, 87, 113, 115, 116
Denmark, 3, 80, 82, 85
reservation, 84, 85
Destabilisation, 23, 34, 50, 189
Development
extractive and colonizing agendas of industrial development, 187
Millennium Development Goals, 55

sustainable development, 38, 55, 62, 63, 65–68, 88, 151–153, 156, 158
Sustainable Development Goals, 56
Dialogue, 4, 50, 70, 72–74, 77, 78, 80, 106, 107, 117, 121–123, 126, 127, 132, 158, 185
cosmopolitan dialogue principle, 62
Diversity, 9, 21, 33, 34, 36, 40, 41, 52, 67, 68, 72, 86, 89, 90, 107, 111, 121, 122, 145, 162, 169, 187, 188
Dystopia, 32

E
Economic Commission for Latin America and the Caribbean (ECLAC), 151, 152
Economy
developed economy, 4, 86
developing economy, 4
economy in transition, 4, 67, 86, 150
English School, 8–10, 19–23, 31, 33, 76, 110, 118, 147, 179, 180, 182
Environmental awareness and education, 6, 130
proposal for an environmental widening participation programme, 130
Environmental communication, 105, 119
self-in-place logic, 119, 123, 126
trinity of voice, 105, 106, 122
Environmental defenders, 12, 30, 118, 154–156, 180, 181
Environmental health, 35, 60, 89, 108, 115
Environmental impact assessment (EIA), 61, 65, 68, 110, 120, 121
Environmental rights, 3, 5–7, 56, 60, 64–66, 84, 85, 88, 114, 127, 131, 132, 141, 152, 181, 182, 184, 185, 187
procedural, 5, 55, 65, 66, 79, 127, 132, 141
substantive, 5, 7, 53, 65, 85, 152, 184
Environmental security, 24
Environment for Europe
first ministerial conference: Dobříš Castle, Czechoslovakia, 78
fourth ministerial conference: Aarhus, Denmark, 3, 80, 82, 83, 85
second ministerial conference: Lucerne, Switzerland, 78
third ministerial conference: Sofia, Bulgaria, 78
Equity, diversity and inclusion (EDI), 111, 186
Escazú Agreement, 152
Ethics
ethical ambition, 10, 19, 38, 39, 142, 169
human-oriented ethics, 9
other-oriented ethics, 33, 39
European Coal and Steel Community, 30
European Community (EC), 61, 80, 86
Directive 85/337, 61
Directive 90/313, 61
European Convention on Human Rights (ECHR), 56, 57
European Court of Human Rights (ECtHR), 55, 56
European Court of Justice (ECJ), 57
European ECO Forum, 148, 153, 155, 158
European Environmental Bureau (EEB), 69, 155

Extinction, 25

F
Faroe Islands, 85
Force, 3, 23–25, 27, 36, 51, 54, 56, 83, 116, 119, 155, 158
Freedom of information, 54, 55

G
G7, 64
Geopolitics, 156, 157, 180, 181
Glasnost, 69, 77, 185
Globalisation, 25, 27, 151
Global justice and injustice, 149, 152, 156
Goldman Environmental Prize, 84
Governance, 4–6, 21, 28, 37, 39, 57, 67, 84, 85, 87, 107, 108, 110, 112, 114, 117, 118, 132, 141, 148, 151, 168, 184, 185
Greenland, 85
Gross domestic product, 86
Guinea-Bissau, 4, 152

H
Heterogeneity, 34
Higher education (HE), 111
  widening participation, 111
Human
  agency, 7, 28, 60, 77, 108, 133
  associations, 2
  care, 27
  condition, 24
  dignity, 54, 57, 88, 122
  emancipation, 40, 141, 182
  empowerment, 1, 6, 12, 31, 39, 41, 59, 60, 62, 69, 75, 88, 105, 110, 111, 113, 114, 116, 117, 119, 121, 125–127, 129, 141, 142, 148, 166, 168, 185
  equality, 108
  health, 112
  legal actorhood, 131, 168
  referent of ethical concern, 4, 34
  rights, 6, 11, 30, 31, 33, 37, 49, 50, 52, 53, 55–57, 59, 64, 76, 88, 108, 122, 148, 153, 155, 180, 183
  subject of international law, 36
Humanitarian intervention, 36
'Humankindness', 9, 12, 41, 63, 89, 147, 156, 159, 168, 169, 185

I
Injunctive relief, 130, 132
  as 'cease and desist' device, 130
  pyrrhic victor, 130, 182
Institutions, 20, 32, 37, 63, 74, 106, 152
Intergovernmental Conference of Experts on the Scientific Basis for Rational Use and Conservation of the Resources of the Biosphere, 58
Intergovernmental Panel on Climate Change, 64
International Bill of Rights, 51, 54
International Covenant on Civil and Political Rights, 52, 53
International Covenant on Economic, Social and Cultural Rights, 52–54
International Criminal Court (ICC), 8, 142, 161, 167
International diplomacy, 84
  of abstinence, 84
International environmental agreements, 1, 2, 8, 10, 20, 32, 35, 41, 60, 65, 81, 132, 143, 144, 157, 159, 179, 188
International interests, 35, 53, 185
International law, 3, 5, 21, 32, 35, 52, 55, 64–66, 74, 82, 84–87,

89, 143, 155, 159, 160, 164, 165, 185, 186
classical understandings, 82
reservation, 3, 186
International order, 2, 9–11, 13, 30, 31, 33, 35, 50–52, 54, 57, 66, 68, 77, 88, 89, 118, 132, 183, 184, 186, 187, 189
as 'safety scaffolding', 68, 88, 187, 189
International relations (IR), 2, 3, 8–10, 12, 19, 20, 22, 23, 25, 26, 29, 34, 39, 40, 51, 68, 70, 72, 89, 113, 114, 179, 180, 183, 186
International society, 2, 5, 8–10, 12, 20–23, 26, 31–33, 35, 39, 40, 58, 68, 88, 89, 106, 169, 183–185, 189
convergence by states upon issues of shared concern, 9, 20, 32
international society arrangements, 20, 21, 26, 35, 89, 131, 169, 182, 184–187
International system, 10, 20, 21, 31
International Technical Conference on the Protection of Nature, 57
International Union for Conservation of Nature (IUCN), 76

J
Jus ad bellum, 51
Justice, 11, 33, 52, 53, 64, 72, 79, 88, 105, 111, 131, 132
Justiciability, 12, 127, 131, 133

K
Kiev Conference on Environmental Protection, 73

L
League of Nations, 51
Legitimacy, 2, 19, 29, 32, 37, 127, 131–133, 142, 169, 184, 188
Leviathan, 23, 186
Liechtenstein, 86
Locus standi, 79, 106, 110, 129, 131, 133, 188
London School of Economics, 19

M
Macht, 24
Man and the Biosphere programme, 58
Masculinity, 25
Millennium Development Goals (MDGs), 55, 152, 153
Monaco, 86
Mongolia, 152
Montreal Protocol on Substances that Deplete the Ozone Layer, 61
Morocco, 152
Multilateral environmental agreements. *See* International environmental agreements
Multilateralism, 3, 8, 10–12, 20, 22, 30, 57, 61, 65, 68, 78, 80, 87, 88, 133, 142, 144, 155–157, 159, 167–169, 180, 183–186, 188

N
National interest, 2, 8, 22–24, 31, 32, 67, 88, 89, 114, 118, 126, 132, 161, 180
Need to know, 113, 117, 132, 185
Nongovernmental organisations (NGO)
Ecoglasnost, 76
observer status, 147
Socio-Ecological Union, 75

Non-intervention, 33, 34, 36, 50, 52, 68
Nonstate actors, 21, 22, 39, 65, 77, 78, 80–83, 85, 87, 89, 107, 147, 158, 167, 168, 185
  parity of actorhood, 80–82, 85, 87, 89, 107, 147, 159, 168, 182

P
Perestroika, 69, 77, 185
Pluralism, 9, 19, 21–23, 33–35, 38–40, 51, 83, 89, 90, 169, 189
  international harm conventions, 33
Population, 2, 23, 78, 85, 89, 114, 118, 132, 161, 184, 187
Post-sovereign arrangements, 26
Power, 9, 11, 12, 23, 24, 26, 27, 30, 37, 39, 40, 50, 59, 60, 69, 70, 72, 83, 89, 90, 106, 108, 111, 117, 118, 120–122, 124, 126, 132, 133, 142, 162, 169, 181–183, 186, 188
  asymmetries and imbalances, 108, 121, 126, 133, 183
  power frameworks, 39
Prefiguration, 21, 27, 28
Private sector, 6, 106, 130
Procedural justice, 107
Public, 4–8, 29, 53–55, 58, 61–64, 66–68, 73–76, 78–82, 85, 105–117, 119, 120, 131, 145, 148–150, 153, 166, 184
  player-partner dual role, 107, 115
Public authorities, 113, 124, 129
Public participation
  arnstein's ladder, 107
  executive regulations and legally binding normative instruments, 109, 119, 124
  gradient of participation, 123, 124, 126
  identification of the public concerned, 122, 126
  plans, programmes and policies, 119, 123, 124
  specific activities, 119

R
Rationalism, 20–22, 179, 184
Realism, 20, 21, 23–26, 30, 31, 33, 41, 70, 179, 180
Responsibility, 2, 65, 67, 78, 83, 110, 116, 119, 124, 131
Responsibility to Protect (R2P), 36
Revolutionism, 20–23, 25–31, 41, 133, 179, 182, 183
Right to know, 63, 113, 114, 117, 118, 132, 185
Rio Declaration. *See* United Nations Conference on Environment and Development (UNCED)
Rio Earth Summit. *See* United Nations Conference on Environment and Development (UNCED)
Russian Federation
  American oil joint venture, 75
  defence ministry, 84
  Federal Security Service, 84
  Kremlin, 71, 73, 84
  Nikitin, Alexander, 84
  Northern Fleet, 84
  refusal to sign Aarhus Convention, 30, 180
  Supreme Arbitration Court, 75
  Supreme Court, 84

S
Scientific Conference on the Conservation and Utilisation of Resources, 57
Securitisation, 24

Security, 24, 84, 88, 113, 114, 118, 123, 126, 132, 133, 161, 180, 186
Social movements, 2, 21
Solidarisation, 8–11, 19, 20, 23, 38, 39, 41, 89, 90, 106, 118, 126, 132, 141, 147, 157, 160, 161, 168, 180, 189
Solidarism, 9, 10, 19, 22, 23, 26, 33, 36–40, 78, 80, 89, 90, 108, 109, 122, 123, 147, 149, 156, 162, 168, 169, 186, 189
  cosmopolitan harm conventions, 33
Sovereignty
  affirmations and iterations of sovereignty, 59
  collective investment of sovereignt, 77, 159, 185
  counterbalancing in favour of sovereignty, 10, 83, 89, 132, 142, 147, 157, 165, 168, 186, 187
  doggedness of sovereignty, 144, 145, 156, 168, 188
  ethical agitation and maturation of sovereignty, 10, 118
  as gatekeeping, 124, 126
  persistence of, and insistence upon, 11, 68, 84, 85, 133, 169, 186
  responsible sovereignty, 12, 22, 37, 189
  sovereign prerogative, 2, 8, 9, 13, 41, 67, 84, 118, 123, 126, 127, 132, 146, 157, 162, 166, 168, 186, 188, 189
  as sovereignty *over*, 124, 162, 167, 187, 188
  tyranny of sovereignty, 118
Spaceship Earth, 27
Stabilisation, 10, 30, 77, 186, 187, 189
States
  ability to enter and withdraw from the diplomatic 'pitch' at will, 5, 34, 161
  conflict, 2, 23
  consent, 2, 5, 8, 10, 11, 20, 31, 40, 60, 64, 68, 69, 77, 81, 83, 89, 106, 132, 141, 143, 156, 157, 161, 165–169, 179, 183, 186, 188
  domestic jurisdictions, 38, 51, 84, 167
  as gatekeepers, 39
  survivalism, 181
  threshold of feasibility and tolerance, 2, 3, 11, 64, 66, 83, 87, 118, 122, 127, 132, 141, 142, 155, 156, 161, 169, 184
  unilateral reticence, 142, 161, 162
Stockholm Conference. *See* United Nations Conference on the Human Environment (UNCHE)
Stockholm Declaration. *See* United Nations Conference on the Human Environment (UNCHE)
Sub-Commission on Prevention of Discrimination and Protection of Minorities, 64
  Draft Principles on Human Rights and the Environment, 64
  Ksentini Report, 64
Superpowers, 23, 49, 69, 70, 72–75, 77
  dialogue and mutual learning, 69, 77
Supranational authority, 23, 184
Sustainable Development Goals (SDGs), 56, 152, 153
System change, 21, 29, 39

T

Territory, 23–26, 70, 72, 77, 129, 181, 183, 188

territorial integrity, 23, 24, 33, 34, 50
territorial trap, 25
Trade, 87, 152
Transformation, 9, 10, 12, 41, 133, 189
post-sovereign, 26, 27, 37, 39, 41, 142, 168, 183, 189
post-Westphalian, 9, 12
Transparency, 3, 6, 30, 63, 71, 75, 79, 105, 111, 114, 115, 149, 152
Trust, 115, 117

U

Ukraine, 146, 157, 181
Chernobyl, 75
Union of Soviet Socialist Republics (USSR), 29, 49, 69–73, 75, 76
All-Union Law on the Order of Appeal to a Court of Illegal Actions of State Bodies and Officials Infringing the Rights of Citizens, 75
Goskompriroda (USSR State Committee for Nature Protection), 75
United Kingdom (UK)
Data Protection Act, 114
Government Security Classifications, 114
Official Secrets Acts, 114
reservation, 87
United Nations Commission on Sustainable Development, 65
United Nations Conference on Environment and Development (UNCED)
Agenda 21, 65, 67
Rio Declaration, 64, 65
Rio Principle 10, 66, 78, 79
United Nations Conference on the Human Environment (UNCHE), 57, 59, 60, 63
Stockholm Declaration, 59, 60
United Nations Economic Commission for Europe (UNECE)
Committee on Environmental Policy, 80
draft convention on environmental information access and public participation, 80
Draft Guidelines on Access to Environmental Information and Public Participation in Environmental Decision-making, 79
Environment and Human Settlements Division, 80
United Nations Environment Programme (UNEP), 60, 61, 76, 151
Bali Guidelines, 151
Goals and Principles on Environmental Impact Assessment, 61
United Nations Framework Convention on Climate Change (UNFCCC), 65
United Nations General Assembly
Declaration on Social Progress and Development, 55
Resolution 59(I), 54
resolution on the need to ensure a healthy environment for the wellbeing of individuals, 64
World Charter for Nature, 61
United Nations Secretary-General, 3, 82, 148, 152
United Nations Security Council, 51

United States of America (US), 49, 70–74, 76, 84, 87, 142, 161, 167, 168
Universal/particular debate, 76
Universal Declaration of Human Rights, 30, 36, 52, 53
University of Oxford, 19
Utopia, 29

V

Violence of silence, 6, 133, 185
Voice
  amplification, 4, 12, 50, 125, 132, 141, 168, 185
  articulation and expression, 182
  hearing and being heard, 62
  reception, inclusion and exclusion, 122
Voluntarism, 20, 22, 34, 64, 84, 155, 168
Voting, 8
Vulnerability
  environmental vulnerability, 10
  human vulnerability, 63
  state vulnerability, 33

W

War, 21, 25, 30, 50, 53, 68, 181, 187
Warsaw Pact, 4, 49, 70, 86, 185
Westphalianism, 24
World Climate Conference, 61
World Commission on Environment and Development (WCED), 38, 62, 63
World government, 11, 19, 21, 23
World Health Organisation, 64
  European Charter on Environment and Health, 64
World society, 20, 21, 26, 76, 77, 110, 182, 183, 185, 187

GPSR Compliance

The European Union's (EU) General Product Safety Regulation (GPSR) is a set of rules that requires consumer products to be safe and our obligations to ensure this.

If you have any concerns about our products, you can contact us on ProductSafety@springernature.com

In case Publisher is established outside the EU, the EU authorized representative is:

Springer Nature Customer Service Center GmbH
Europaplatz 3
69115 Heidelberg, Germany

Kolibri 360 GmbH
Ferdinand-Jühlke-Straße 7
99095 Erfurt, Deutschland
produktsicherheit@kolibri360.de